Techniques for the Rapid Detection of Plant Pathogens

Techniques for the

Rapid Detection

of Plant Pathogens

Edited by J.M. Duncan & L. Torrance
Scottish Crop Research Institute, Invergowrie, Dundee, Scotland

With an Introduction by I.M. Smith
Director-General, European and Mediterranean
Plant Protection Organization, Paris

PUBLISHED FOR THE

BRITISH SOCIETY FOR PLANT PATHOLOGY BY

BLACKWELL SCIENTIFIC PUBLICATIONS

OXFORD LONDON EDINBURGH

BOSTON MELBOURNE PARIS BERLIN VIENNA

© 1992 by
Blackwell Scientific Publications
Editorial offices:
Osney Mead, Oxford OX2 0EL
25 John Street, London WC1N 2BL
23 Ainslie Place, Edinburgh EH3 6AJ
3 Cambridge Center, Cambridge
  Massachusetts, 02142, USA
54 University Street, Carlton
  Victoria 3053, Australia

Other Editorial offices:
Arnette SA
2, rue Casimir-Delavigne
75006 Paris
France

Blackwell Wissenschaft
Meinekestrasse 4
D-1000 Berlin 15
Germany

Blackwell MZV
Feldgasse 13
A-1238 Wien
Austria

First published 1992

Set by Setrite Typesetters, Hong Kong
Printed and bound in Great Britain
by The University Press, Cambridge

DISTRIBUTORS

  Marston Book Services Ltd
  PO Box 87
  Oxford OX2 0DT
  (Orders: Tel: 0865 791155
          Fax: 0865 791927
          Telex: 837515)

USA
  Blackwell Scientific Publications, Inc.
  3 Cambridge Center
  Cambridge, MA 02142
  (Orders: Tel: (800) 759−6102)

Canada
  Oxford University Press
  70 Wynford Drive
  Don Mills
  Ontario M3C 1J9
  (Orders: Tel: (416) 441−2941)

Australia
  Blackwell Scientific Publications
  (Australia) Pty Ltd
  54 University Street
  Carlton, Victoria 3053
  (Orders: Tel: (03) 347−0300)

British Library
Cataloguing in Publication Data

Techniques for the rapid detection
of plant pathogens.
  1. Plants. Pathogens
  I. Duncan, J.M., II. Torrance, L.
  III. British Society for Plant Pathology
  581.23

  ISBN 0−632−03066−6

Library of Congress
Cataloging-in-Publication Data

Techniques for the rapid detection
of plant pathogens
    edited by J.M. Duncan & L. Torrance.
        p.   cm.
    ISBN 0−632−03066−6
  1. Plant diseases−Diagnosis.
  2. Phytopathogenic microorganisms−
  Detection.
  I. Duncan, J.M. II. Torrance, L.
  SB731.T43   1992
  632−dc20

Contents

Contributors

S. BALL *Official Seed Testing Station, National Institute of Agricultural Botany, Cambridge CB3 0LE, UK*

A.A.G. CANDLISH *Rhône-Poulenc Diagnostics Limited, Montrose House, 187 George Street, Glasgow G1 1YT, UK*

A.C. CASSELLS *Department of Plant Science and Biotechnology, (UCC) Ltd, University College, Cork, Ireland*

M.F. CLARK *Horticultural Research International, East Malling, Maidstone, Kent ME19 6BJ, UK*

A. CODDINGTON *School of Biological Sciences, University of East Anglia, Norwich NR4 7TJ, UK*

F.M. DEWEY *Department of Plant Sciences, University of Oxford, South Parks Road, Oxford OX1 3RB, UK*

D.S. GOULD *School of Biological Sciences, University of East Anglia, Norwich NR4 7TJ, UK*

G.D. GROTHAUS *Agri-Diagnostics Associates, 2611 Branch Pike, Cinnaminson, NJ 08077, USA*

P. GUGERLI *Federal Agricultural Research Station of Changins, CH-1260 Nyon, Switzerland*

A.T. JONES *Virology Department, Scottish Crop Research Institute, Invergowrie, Dundee DD2 5DA, UK*

S.A. MILLER *Agri-Diagnostics Associates, 2611 Branch Pike, Cinnaminson, NJ 08077, USA*

F.P. PETERSEN *Agri-Diagnostics Associates, 2611 Branch Pike, Cinnaminson, NJ 08077, USA*

M. QUERCI *The International Potato Center, P.O. Box 5969, Lima, Peru*

J. REEVES *Official Seed Testing Station, National Institute of Agricultural Botany, Cambridge CB3 0LE, UK*

J.H. RITTENBURG *Agri-Diagnostics Associates, 2611 Branch Pike, Cinnaminson, NJ 08077, USA*

L.F. SALAZAR *The International Potato Center, P.O. Box 5969, Lima, Peru*

I.M. SMITH *Director-General, European and Mediterranean Plant Protection Organization, 1 Rue Le Nôtre, 75016 Paris, France*

J.E. SMITH *Division of Applied Microbiology, Department of Bioscience & Biotechnology, University of Strathclyde, Glasgow G4 0NS, UK*

D.E. STEAD *Ministry of Agriculture, Fisheries and Food, Central Science Laboratory, Hatching Green, Harpenden, Herts AL5 2BD, UK*

W.H. STIMSON *Division of Immunology, Department of Bioscience & Biotechnology, University of Strathclyde, Glasgow G4 0NS, UK*

L. TORRANCE *Virology Department, Scottish Crop Research Institute, Invergowrie, Dundee DD2 5DA, UK*

A. VIVIAN *Science Department, Bristol Polytechnic, Coldharbour Lane, Frenchay, Bristol BS16 1QY, UK*

Preface

In the production of crops, whether in intensive systems or otherwise, one of the most important inputs must be the plant material itself. While plant breeders have sought to improve the quality and yield of plants through selective breeding, the traditional role of the pathologist has been to ensure that the planting material, whether true seed or vegetative propagule, is free of damaging diseases. This role is of increasing importance as the scale of movement of plant material within and between countries increases.

With the advent of new techniques in biology comes new opportunities. Breeders can use the new biotechnology to locate and manipulate important genes in the plant, whilst the pathologists can use it to detect the presence of pathogens in host plant tissue, to identify the pathogens and to diagnose diseases of a crop with rapidity, accuracy and sensitivity.

Increasing interest in the application of new, rapid techniques for the detection of plant pathogens led, in 1989, to a conference on this theme being held by the British Society for Plant Pathology, with the support of the British Crop Protection Council, at the University of East Anglia in Norwich. This book has its origins in that meeting, but it is not a proceedings of it. Instead, it is intended more as a 'methods' book which gives practical details of how to apply a range of the newer, rapid techniques to a wide range of diseases and problems. To this end most chapters have extensive appendices which can be used as laboratory protocols by those who want to try out the techniques.

The commonest approaches to the detection of pathogens and the diagnosis of disease involve the use of immunology and nucleic acid technology, and two of the three sections in the book are devoted to techniques employing these technologies. A part of the immunology section — that on detecting and identifying bacteria — covers other technologies which have not yet been used widely but which merit consideration by pathologists interested in this field.

The third section of the book deals with the practical application of the techniques described in the first two sections. As the chapter by Gugerli illustrates, there is considerable commercial potential in the field of rapid detection and diagnosis of viruses. This potential extends to diseases caused by fungi and bacteria and those involved in early diagnosis should be aware of it and of some of the problems which they face if they are to bring their work to the stage of commercial exploitation.

It is our hope that this book will be of value to students, scientists and others interested in the rapid detection and diagnosis of plant disease and that it will inspire some to employ the new technologies to tackle some of their problems.

The names of the suppliers given in the text are examples based on the experience of the authors. They are not intended to represent endorsements of the products or to be a definitive listing of suppliers.

J.M. Duncan and L. Torrance
Scottish Crop Research Institute

List of abbreviations

A	Absorbance
AAT	Adenine, aminopterin, thymidine
AF	Aflatoxin
APRT	Adenosine phosphoribosyl transferase
ASBVd	Avocado sunblotch viroid
ATA	Antibody trapped antigen
BCIP	5-bromo-4-chloro-3-indolyl phosphate p-toluidine salt
Bis	N,N'-methylene-bis-acrylamide
bp	Base pairs
BSA	Bovine serum albumin
BYDV	Barley yellow dwarf virus
CDAZ	Czapek Dox plus AZ liquid medium
cDNA	Complementary DNA
CFA	Complete Freund's adjuvant
CFU	Colony forming units
CIP	International Potato Center
CM	Conditioned medium supplement
CMV	Cucumber mosaic virus
CNV	Cucumber necrosis virus
conc.	Concentrated/concentration
c.p.m.	Counts per minute
CSF	Cell culture supernatant fluids
CTV	Citrus tristeza virus
dabELISA	Double-antibody ELISA
DAPI	4',6-diamidino-2-phenylindole
DEPC	Diethyl pyrocarbonate
DIECA	Sodium diethyldithiocarbamate
DMSO	Dimethyl sulphoxide
DNA	Deoxyribonucleic acid
DNase	Deoxyribonuclease
dNTP	Deoxynucleoside triphosphate
ds	Double-stranded
DTT	Dithiothreitol
ECL	Equivalent chain length
EDTA	Ethylenediaminetetraacetic acid disodium salt

ELISA	Enzyme-linked immunosorbent assay
FACS	Fluorescence-activated cell sorter
FAME	Fatty acid methyl esters
FCS	Foetal calf serum
FID	Flame ionization detection
FITC	Fluorescein isothiocyanate
G	L-glutamine
GC	Gas chromatography
HAT	Hypoxanthine, aminopterin, thymidine
HAZA	Hypoxanthine, azaserine
HPLC	High performance liquid chromatography
HRPT	Hypoxanthine guanine phosphoribosyl transferase
IEF	Isoelectric focussing
IF	Immunofluorescence
IFA	Incomplete Freund's adjuvant
Ig	Immunoglobulin
i.p.	Intraperitoneal
IPM	Integrated pest management
ISEM	Immunosorbent electron microscopy
i.v.	Intravenous
kb	Kilobase
KLH	Keyhole limpet haemocyanin
LMV	Lettuce mosaic potyvirus
MAbs	Monoclonal antibodies
MLO	Mycoplasma-like organisms
Mol. wt	Molecular weight
NaPB	Sodium phosphate buffer
NASH	Nucleic acid spot hybridization
NBT	p-Nitro blue tetrazolium chloride
OD	Optical density
OTA	Ochratoxin A
PAGE	Polyacrylamide gel electrophoresis
PBS	Phosphate buffered saline
PCR	Polymerase chain reaction
PDA	Potato dextrose agar
PEG	Polyethylene glycol
PLRV	Potato leafroll virus
PMSF	Phenylmethylsulphonyl fluoride
PSF	Penicillin, streptomycin, fungizone
PSTVd	Potato spindle tuber viroid
PTA	Plate trapped antigen
PVDF	Polyvinylidene difluoride
PVP	Polyvinylpyrrolidone
PVX	Potato virus X
RF	Replicative form
RFLP	Restriction fragment length polymorphism

RI	Replicative intermediate
RIA	Radioimmunoassay
RNA	Ribonucleic acid
RNase	Ribonuclease
SA–AP	Streptavidin–alkaline phosphatase conjugate
SDS	Sodium dodecyl sulphate
SDW	Sterile distilled water
ss	Single-stranded
SSC	Saline sodium citrate
STE	Saline, Tris, EDTA
TAE	Tris, acetic acid, EDTA
TBE	Tris, boric acid, EDTA
TBS	Tris buffered saline
TBSV	Tomato bushy stunt virus
TCA	Trichloroacetic acid
TE	Tris, EDTA
TEMED	N,N,N',N'-tetramethylethylenediamine
TLC	Thin layer chromatography
TMB	3,3',5,5'-tetramethylbenzidine
TMS	Tris, magnesium, saline
TMV	Tobacco mosaic virus
TNV	Tobacco necrosis virus
TPE	Tris, phosphate, EDTA
TPS	True potato seed
Tris	Tris(hydroxymethyl)aminomethane
TRV	Tobacco rattle virus
TSA	Trypticase soy agar
Tween	Polyoxyethylenesorbitan (Tween 20)
u.v.	Ultraviolet
VCV	Vicia cryptic virus

Introduction: practical implications of the development of new techniques

The last 5–10 years have seen remarkable progress in the development of new techniques for detection and diagnosis of plant pathogens. This is to a certain extent, of course, the consequence of developments made in the medical field, but the plant pathological sector has now acquired its own momentum as the range of practical applications continues to expand.

Detection in planting material

The detection of pathogens in planting material is a fundamental element in the production of what is known as certified stock. All planting material should, in any case, meet reasonable health standards, but for some critical pathogens it is worthwhile to ensure that these standards are officially set by compulsory or optional certification schemes.

Certain pathogens, mainly bacteria and fungi but also a few viruses, are principally transmitted by seed. Their control is essentially through the production of seeds in which the pathogens are absent or do not exceed a defined low threshold of contamination. Traditional seed-testing methods depend on isolation *in vitro* or on bioassays, which are relatively slow and laborious. Perhaps partly because of difficulties of reproducibility of results, such test methods have been applied relatively little in seed certification schemes which concern only a few specific pathogens. The new techniques, particularly serological, offer the prospect of being simpler, more rapid, and more sensitive. As a consequence, it seems probable that they will enable seed to be more extensively certified for the presence or absence of pathogens. It is true that many seed-borne pathogens are fungi, the group to which the new techniques have up till now been least applied, but, as we shall see below, this may only be a temporary situation.

Even wider possibilities arise with pathogens transmitted with vegetatively propagated planting material: potatoes, fruit crops and many ornamentals. For seed potatoes official phytosanitary certification schemes are already compulsory practically everywhere, while for the other crops they are compulsory

1

in some countries (e.g. The Netherlands, Denmark) and are optionally open to growers in other countries. Viruses are the main pathogens for which the new techniques offer possibilities. They have already become routine in many cases (potato viruses, plum pox potyvirus, nepoviruses) and are likely to be more and more widely used. It then becomes more practical to use similar techniques for other pathogens and we already have the case in France where, in pelargonium certification, enzyme-linked immunosorbent assay (ELISA) testing for tomato ringspot nepovirus will be accompanied by new ELISA tests for *Xanthomonas campestris* pv. *pelargonii* and for *Verticillium albo-atrum*.

In fact, in the Member States of the European Community, phytosanitary certification of vegetatively propagated planting material is likely, by 1993, to become compulsory for many classes of crops, and this will greatly increase the sheer volume of tests which will have to be performed. Rapid, reproducible and simple tests will become urgently necessary.

Monitoring for disease eradication
Linked to the need to test planting material is a general need to monitor planted material for the same pathogens. Many countries aim to eradicate certain virus or bacterial diseases, at least from commercial production (e.g. citrus tristeza closterovirus, plum pox potyvirus and *Clavibacter michiganensis* subsp. *sepedonicus*). Planted material has to be surveyed, infected plants destroyed or infested fields taken out of production. Such systematic surveying also calls for sensitive, specific and rapid techniques.

Diagnostics
Diagnosis as such is a practical application which existed long before these new techniques were introduced. Certainly expert diagnostic laboratories, whether in the private or the public sector, will be able to improve their performance, reliability, speed and sensitivity by calling on new techniques. The users of their services will probably benefit from greater availability and lower cost but the nature of the service will probably not undergo a qualitative change, at least until some of the research implications discussed below have extended into practical use.

One area where there is a possibility of a quite new approach is in the marketing of kits for non-experts and it will be interesting to see how this develops. Kits exist for certain very specific applications (e.g. potato viruses) but these still target material in certification schemes. The new areas would be kits for the extension officer, for the grower, or even for the amateur gardener. Despite some claims it is not clear that this is a rapidly developing market. We have heard of the kit for turf-grass diseases marketed in the USA (Chapter 12), which is, no doubt, excellent in itself. However, agricultural pathologists will no doubt be slightly surprised that such a relatively marginal application should occupy the centre of the stage. Where are the kits for the cereal grower or for the glasshouse tomato grower? The real markets perhaps do not coincide with pathologists' preconceptions.

In particular, it is interesting to explore whether the new techniques could aid not only diagnosis but also forecasting and warning services and, in consequence, integrated pest management (IPM) systems. Though this appears as a very attractive potential market it is doubtful whether it can really be developed. Warning systems for epidemic diseases, if they include any direct pest monitoring at all, mostly depend on very early detection of rare infection foci: the difficulty is in adequate sampling rather than sensitivity or reliability, so new techniques perhaps have little to offer. For soil-borne pathogens, with a slower rate of multiplication and spread, one can envisage soil tests to determine whether a crop should be planted on a plot outdoors or under glass. The question is then whether the very small concentrations of pathogens in soil can be better detected by the new techniques than by the classical isolation or bioassay techniques.

Some research implications

On the research side, we can mention at least two areas of interest although there are certainly others (depending on one's own research orientation). Bacteriologists, and especially mycologists, have not yet fully exploited the new techniques on the grounds that they already possess adequate diagnostic criteria, which are indeed closely integrated with the whole basis of the taxonomy of these organisms (biochemical tests for bacteria and morphological characterization of fungi). However, the new methods open wide possibilities for distinguishing taxa which are genuine biological entities but which cannot be distinguished by the classical criteria. Bacterial pathovars may again become proper species, and fungal formae speciales may be properly characterized, classified and synonymized. Difficult genera like *Phytophthora* and *Fusarium* may be more meaningfully classified. It is true that present results show some difficulty in operating at or below the species level: the greater complexity of the fungal and bacterial genomes and greater diversity of potentially specific macromolecules make it harder to isolate the relevant specific elements. However, these are still early results and one can hope for much progress. At a higher taxonomic level relationships between genera and families of fungi, connections between teleomorphs and anamorphs, and the true relationships of imperfect fungi may be better elucidated by the new techniques.

Another research area is linked to cellular and subcellular studies of host–parasite relationships. Techniques which have up till now been mainly applied in electron microscopy of viruses should also find applications for fungi and bacteria, once the specific diagnostic molecules have been made available.

Problems of the new techniques

Finally, it is useful to mention a few problems. The sensitivity of new techniques may allow detection of pathogens where it was previously impossible. In tests on planting material this is theoretically an advance, especially for organisms for which a 'zero tolerance' is required (e.g. quarantine pathogens). However, it is not so certain that the extra rigour obtained is always needed for the purpose; the reality is that sampling methods are designed to ensure an adequate compromise between the labour involved and the undesirable consequences of a false negative result. If new techniques are more sensitive, then sampling procedures can be rethought, rather than simply setting new and much tougher standards. Another problem is that as the materials for the new techniques become widely available their quality must be guaranteed. Already, different sources of antisera are recognized to vary in quality. As the volume increases and the production process moves away from the laboratories, which are also actual users of the materials, there will be a need for accepted quality standards.

Conclusion

The British Society for Plant Pathology (BSPP) is to be congratulated for organizing the conference at Norwich at which these important new developments were reviewed, and is again to be congratulated for deciding to produce a book based on and developed from this conference. It is a subject of great theoretical and practical interest and I believe that almost all plant pathologists will see their working practices eventually affected by these new developments.

Section 1
Immunological Techniques

1 Serological methods to detect plant viruses: production and use of monoclonal antibodies

L. TORRANCE

Introduction

Serological methods have been used to detect and identify plant viruses for many years (Ball, 1974; Torrance & Jones, 1981; van Regenmortel, 1982). The early tests relied on the ability of specific antibodies to combine with virus particles to give a visible precipitate, such as a line in an agar gel or a floccular mass in liquid. Also, by mixing drops of infective plant sap and antibodies on a slide, components of sap such as chloroplasts became enmeshed in the aggregates of virus particles and antibodies to give a visible agglutination reaction (chloroplast agglutination). These methods were adequate for tests on herbaceous indicator plants or partially purified virus preparations but they required large amounts of antiserum and were rather insensitive.

In a second generation of tests, the antibody−antigen reaction was amplified to increase the sensitivity of virus detection. For example, the latex test (Bercks, 1967; Abu Salih et al., 1968) involved coating the surface of latex beads with antibodies by passive adsorption, then mixing drops of the sensitized latex with drops of plant sap as in the chloroplast agglutination test. It was more sensitive and used less antiserum than the other tests, and was very quick and simple, but non-specific reactions sometimes occurred with plant sap.

A major advance in terms of sensitivity and the types of plant material which could be tested came with the introduction of enzyme-labelled antibodies. The enzyme-linked immunosorbent assay (ELISA) was introduced to plant virology as an antibody sandwich assay where virus-specific antibodies were linked to the enzyme alkaline phosphatase (Voller et al., 1976; Clark & Adams, 1977). It was rapidly shown to be an extremely sensitive method and could be used to detect viruses in previously intractable tissues such as the flesh of potato tubers, leaves of tree fruit (apple, plum and citrus), and the bulb flesh of narcissus, tulip and other bulbous ornamentals (Torrance & Jones, 1981; Cooper & Edwards, 1986). The other great advantage of ELISA is the potential for testing very large numbers of samples. Using ELISA together with improved devices for sample extraction, automatic plate washers and reagent dispensers, the Dutch bulb inspection service can process 1000 samples per hour and more than 2 000 000 tulip and lily bulbs are screened routinely during the winter months for tulip breaking and lily X viruses (A.R. van Schadewijk, pers. comm.). Also, the British Ministry of Agriculture, Fisheries and Food Central Science Laboratory at Harpenden has tested up to 8000 plum leaves per year in a mass screening programme for plum pox virus in English plum nurseries (Torrance & Jones, 1981). Since the introduction of the direct double-antibody sandwich ELISA there has been a proliferation of methods based on enzyme-labelled antibodies: indirect methods using anti-rabbit antibodies linked to enzyme; use of many different types of enzyme; different substrates; and use of different supports such as nitrocellulose instead of plastic microtitre plates (Cooper & Edwards, 1986; Barbara & Clark, 1986; Nakamura et al., 1986).

In addition to ELISA, methods have been developed using the electron microscope to visualize the antigen−antibody reaction (Roberts, 1985) and recently these methods have been enhanced by the incorporation of gold-labelled antibodies (van Lent & Verduin, 1986). The electron microscope techniques can be useful for tests on small numbers of samples.

Taken together, modern serological techniques for detecting and identifying plant viruses have revolutionized the possibilities for the production of large quantities of virus-tested stocks of many crops. They have also improved the quality and speed of service offered by extension and quarantine authorities.

Production of polyclonal antisera

The immune response is the result of a complex series of processes in the animal and for a detailed description the reader is referred to Campbell (1984), Kimball (1986), Weir et al. (1986) or Harlow & Lane (1988). The response of experimental animals (even inbred species) to immunization with antigen is very variable (Dresser, 1986). Many different factors affect the quality of response obtained and they have not all been studied systematically. Some general rules for viral protein antigens can be found in the literature. The animals should be young healthy adults (>6 weeks old for mice and >12 weeks old for rabbits). Several individuals should be immunized, at least two rabbits or four to six mice (for hybridoma work). Inclusion of Freund's adjuvant with antigen in a water-in-oil emulsion (Appendix 1) is recommended to potentiate the immune response and as a depot for slow release of antigen. The amount injected should be as small as practicable to minimize unwanted reactions to contaminants or minor components and induction of tolerance. Insoluble or aggregated polymeric forms of antigen are said to be more immunogenic (Dresser, 1986; Sela, 1986).

Many different schedules to elicit antibodies to plant virus antigens have been published and most give satisfactory results. They are generally combinations of intravenous and intramuscular injections given to rabbits over a period of several weeks or months. In a study using potexvirus antigens, Koenig & Bercks (1968) obtained a good response which was sustained over 5−6 months with a schedule of two intramuscular injections spaced 1 week apart, the first using complete Freund's adjuvant (CFA) and the second using incomplete Freund's adjuvant (IFA). The animals were bled 2 weeks after the second injection and every 2 weeks subsequently. Van Regenmortel (1982) routinely uses a series of intramuscular injections of antigens emulsified with IFA given at 2-week intervals. I have obtained satisfactory antisera against viruses in at least six different taxonomic groups by giving two sets of intramuscular injections 1−2 weeks apart, the first of immunogen in CFA and the second in IFA (50−500 µg of virus in each injection). For each injection the emulsion was deposited in the thigh muscle, half in each leg.

For weak antigens, Dresser (1986) recommends giving an intramuscular injection of 100−150 µg antigen emulsified in CFA (left thigh); 7−10 days later the same quantity in CFA in the right thigh; rest the animal for 8 weeks; then boost with the antigen in IFA by subcutaneous injection at five to six sites. Bleed 6−8 days later and every 2 weeks until the titre decreases. Some plant virus antisera have been raised in hens using similar injection schedules as for rabbits, and the eggs of immunized laying hens are a rich source of antibodies (van Regenmortel, 1982).

There are occasions when material other than intact virus coat protein is used to elicit antibodies, for example, non-structural potyviral protein (Mowat et al., 1989). Protein bands separated from a mixture by electrophoresis in polyacrylamide gels are electro-eluted and emulsified in Freund's adjuvant, then injected intramuscularly to rabbits and mice (Dresser, 1986; Goding, 1986; Harlow & Lane, 1988). Proteins transferred to nitrocellulose can also be injected after grinding the membrane to a powder (Harlow & Lane, 1988). Viral gene products in the form of fusion proteins generated by molecular cloning techniques, and short synthetic peptides are used to raise antisera (van Regenmortel et al., 1988; Harlow & Lane, 1988). Bahner et al. (1990) produced an antiserum to a fusion protein comprising part of the P5 protein of potato leafroll virus linked to protein A (titre >1/500 by immunoblot assay) by injecting

250 μg in CFA intramuscularly and then, 1 month later, giving three further injections at 20-day intervals of 250 μg in IFA. Antisera to synthetic peptides comprising 10–18 amino acid residues (from a region of the sequence of poliovirus 3 VP1) which had high ELISA titres (>1/10 000) against the peptides were obtained by injecting rabbits intramuscularly with 250–500 μg peptide in CFA; this was repeated after 3 weeks and again after a further 2 weeks, this time without adjuvant. Although the ELISA titres against the peptides were high, the best neutralizing antibody titres (anti-viral titres) were obtained when the peptides were coupled to a carrier protein (keyhole limpet haemocyanin [KLH] or bovine thyroglobulin) or when they were treated with glutaraldehyde (Ferguson et al., 1985).

Peptides of 2000 Da or less should be conjugated to a carrier molecule such as KLH or bovine serum albumin (BSA) to stimulate a better response (Sela, 1986; Makela & Seppala, 1986; Harlow & Lane, 1988). A discussion of immunization protocols for synthetic peptides with several examples can be found in van Regenmortel et al. (1988).

Production of monoclonal antibodies

Monoclonal antibodies (MAbs) against plant viruses were first produced in the early 1980s (Dietzgen & Sander, 1982; Briand et al., 1982; Gugerli & Fries, 1983; Halk et al., 1984; Torrance et al., 1986 a,b), and MAbs have been produced to members of all of the major groups of plant viruses (Hsu, 1984; van Regenmortel, 1986). As with all new technology, some practical experience of MAbs has been necessary to evaluate the advantages and disadvantages of incorporating them in diagnostic assays. For example, it was not fully appreciated that MAbs are epitope specific rather than virus specific (van Regenmortel, 1986), and that epitopes vary in their suitability as targets for detection in routine diagnosis of virus diseases. In addition, the antibody molecules themselves have different properties. The rest of this chapter describes methods used to produce MAbs to plant viruses and some points to consider when using them to devise diagnostic tests.

Techniques for the production of MAbs enable the isolation and culture of individual antibody-secreting B cells from the mixture of B cells (each secreting an antibody with distinct properties) present in the spleen of the immunized animal (Fig. 1.1). The methods described in the appendices are essentially those of Galfrè & Milstein (1981). They have been adapted and modified through time and have given good results with many different plant viruses. In addition to the laboratory protocols given here, the reader is also recommended to refer to

Spleen cells from
immunized mouse/rat ————————————— Myeloma cells
HPRT$^+$ HPRT$^-$

Fuse with polyethylene glycol

Only fused cells grow on selective medium

Assay supernatants for antibody activity
after 1–2 weeks

Expand and then assay cultures for antibody activity

Freeze   Clone   Freeze   Clone   Freeze   Clone

Assay supernatants

Select 2–3 wells to expand and then assay

Freeze 9–12 vials   Clone

Assay supernatants

Select 2–3 wells to expand and then assay

Freeze 12 vials   Grow in flasks

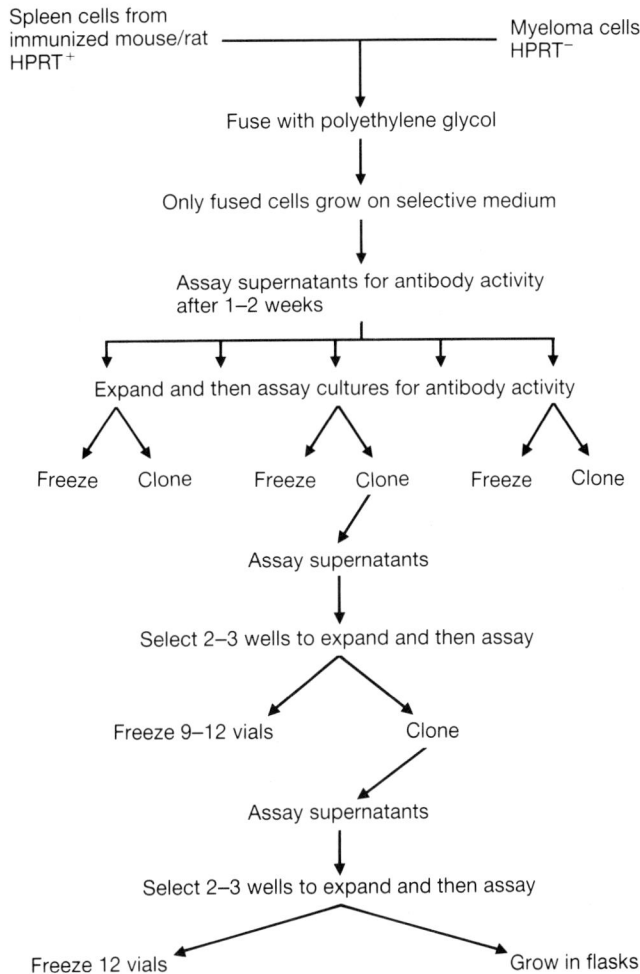

Fig. 1.1 Flow diagram of monoclonal antibody production.

Hudson & Hay (1980), Campbell (1984), Goding (1986), Weir *et al.* (1986) and Harlow & Lane (1988) for general information and many details of MAb production, types of assay and basic immunology, and to Adams (1981) and Freshney (1987) for aspects of cell culture and aseptic technique. Useful information can also be found in suppliers' catalogues (e.g. Sigma, Pierce & Warriner (UK) Ltd.

Assays of cell culture supernatant fluid (CSF)   The first stage in the production of MAbs is to devise one or more assays to detect the antibodies secreted by the fused spleen and myeloma cells (hybridomas). Ideally these assays should give results fast and be capable of screening large numbers of samples. Usually some form of ELISA is preferred. I

employ two different forms of ELISA, either where purified virus particles are adsorbed to the surface of microtitre plates (plate trapped antigen [PTA], Appendix 2) or where a layer of antibodies (usually rabbit polyclonal) are adsorbed to the plates which in turn bind virus particles (antibody trapped antigen [ATA], Appendix 2). These methods obviously detect virus-specific antibodies but it is also possible to screen the CSF merely for the presence or absence of antibody (Appendix 3). This may be useful if the specific assay is not readily adapted to test large numbers.

It is possible to select antibodies which react with a particular protein in a mixture by the immunoblotting technique (Appendix 4). The proteins are separated by electrophoresis in a polyacrylamide gel and then transferred electrophoretically (blotted) to nitrocellulose membrane or other suitable support material. Several gels can be run and blotted in advance and the membranes wrapped in cling film and stored frozen. Monoclonal antibodies are then incubated with the membrane and bind to the transferred antigen, and the bound MAbs are detected by successive incubations with an anti-mouse (or anti-rat) alkaline phosphatase conjugate and then with the enzyme substrate. The product of the enzyme reaction is insoluble and a black/blue precipitate forms at the site of bound enzyme. A multichannel apparatus (the miniblotter range, Biometra; Cross Blot, Sebia) can be used to compare 20−40 samples of CSF at one time (Fig. 1.2).

It is an advantage to know the possible forms of antigen which might be detected in the assay. For example, with viruses which have relatively stable particles such as luteoviruses, the ATA ELISA will be more likely to select antibodies which react with the assembled or relatively undegraded virus particles.

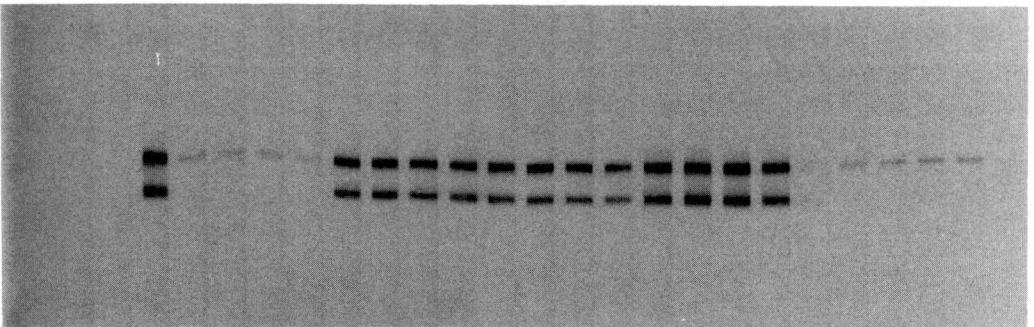

Fig. 1.2 Comparison of immunoblotting reactions of 26 hybridoma cell culture supernatant fluids using multichannel apparatus (Biometra).

These epitopes may be neotopes (epitopes not present on individual protein subunits; van Regenmortel, 1982, 1989) and they may be discontinuous (sometimes called conformation-dependent or assembled topographic) epitopes which are composed of amino acid residues brought into close proximity by the folding of the polypeptide chains (van Regenmortel & Neurath, 1985). However, the particles of potato leafroll luteovirus can be disrupted by dilution in carbonate buffer, pH 9.6 (Massalski & Harrison, 1987) as when preparations are used to coat microtitre plates for PTA ELISA. PTA ELISA and the immunoblotting technique will be more likely to select antibodies which react with epitopes which only become exposed after fragmentation or depolymerization of the capsid (cryptotopes; van Regenmortel, 1989), and they may be continuous epitopes, which are composed of amino acids that occur consecutively in the primary amino acid sequence (van Regenmortel & Neurath, 1985).

The type of epitope detected by the MAb will influence the diagnostic assay which is eventually adopted and it is important to ensure that the screening method is likely to select antibodies with the required reactivity. For example, if an epitope common to a wide range of virus strains is desired then several different strains should be used in the assays. A method for producing potyvirus MAbs was devised which exploited the finding that much of the serological specificity resides in the exposed N-terminal amino acids of the coat protein subunits of potyviruses (Shukla et al., 1988). Virus-specific and potyvirus group cross-reacting MAbs were identified by screening the CSF against complete virus particle proteins and those in which the N-terminal part of the subunit was removed (Shukla & Ward, 1989). However, exposed N- or C-terminal epitopes can be removed by trypsin or other treatments (Koenig & Torrance, 1986; Mackenzie & Tremaine, 1986; Shukla & Ward, 1989) and MAbs specific for these epitopes may not be suitable for use in routine assays where samples are exposed to proteolytic enzymes in plant saps.

There are many different types of screening assays to detect antibody-secreting hybridomas and the examples given above are suited to plant virus work. However, there are others which may be more suited to isolating an antibody with both particular physico-chemical properties (low affinity, effector function, stability) and binding specificity. Kipps & Herzenberg (1986) describe applications of the fluorescence-activated cell sorter (FACS) to detect hybridomas producing MAbs to cell surface antigen. Many other assays are described in Hudson & Hay (1980), Weir et al. (1986) and Harlow & Lane (1988).

Immunization of animals

In immunization protocols for hybridoma work usually 10–100 µg of virus is used per injection, and mice are injected by intra-peritoneal or subcutaneous routes instead of intramuscular routes. All methods incorporate a final booster injection (usually intravenous) approximately 3 days before the fusion. The strain of mouse used is generally BALB/c because the mouse myeloma fusion partners were of BALB/c origin, and so the resultant hybridomas are compatible for ascites production in BALB/c mice (Goding, 1986; Bartal & Hirshaut, 1987; Harlow & Lane, 1988). New Zealand Black-BALB/c F1 hybrid mice have been used for immunization with fungal mycotoxins (Candlish et al., this volume) and may be useful where a poor response is obtained in BALB/c mice (Campbell, 1984). Rats and rat myelomas have also been used very successfully to produce MAbs (Galfrè & Milstein, 1981; Galfrè & Butcher, 1986; Torrance et al., 1986 a,b).

A typical schedule for injecting mice with plant virus antigens would be one intraperitoneal (or a combination of intraperitoneal and subcutaneous) injection of 10–50 µg antigen in CFA, and a second injection 2–4 weeks later with the same amount of antigen in IFA. After a further 2 weeks mice are bled to check the titre. A further 4 (or more) weeks rest period is then allowed before giving a booster injection (50 µg in saline) intravenously in the tail. The fusion is done 3 days later.

Monoclonal antibodies to potato virus X were obtained by immunizing rats as follows: day 0, 100 µg of virus in CFA intramuscularly; day 27, 100 µg in IFA subcutaneously; day 52, 75 µg in saline intravenously. The fusion was done 3 days later (Torrance et al., 1986a). For more information on immunization protocols see Campbell (1984), Eshhar (1985) and Harlow & Lane (1988).

Even though serum titres may be high, this is not a sure sign that the fusion will produce many hybridomas. One of the keys to successful fusion seems to be to induce large numbers of antigen-stimulated B cells in the spleen (Eshhar, 1985; Campbell, 1984; Westerwoudt, 1986). By giving a series of injections (intraperitoneal and/or intravenous) each day for 4 days prior to the fusion instead of the usual one injection 3 days beforehand, Stahli et al. (1980) obtained many more clones secreting specific antibodies to the soluble antigen human chorionic gonadotropin. However, Hudson & Hay (1980) found that a very short schedule was effective in production of hybridomas against another soluble antigen, human IgG. The protocol was simply one intraperitoneal injection of 100 µg IgG in CFA followed 7 days later by a booster injection of 100 µg given intravenously. The cells were fused 4 days later (Hudson & Hay,

1980). This result illustrates the variability of response which can be obtained and shows that it is difficult to make generalizations about injection schedules. The best policy is to make two or three injections to a group of animals and take test bleeds to check the titre. If the response is very poor, a different breed or strain should then be tried.

Alternatives to the above immunization schedules include a way of increasing the yield of antigen-specific hybridomas devised by Fox *et al.* (1981). They injected spleen cells from immunized mice intravenously into syngeneic recipient mice which had been irradiated. The recipients were also injected intraperitoneally with antigen and the fusion was done 3–4 days later. This method is said to increase the proportion of antibody-producing cells over non-specific cells.

Another method which can be employed when very small amounts of antigen are available is to inject antigen directly into the spleen (Spitz, 1986; Spitz & Spitz, 1987). Antibodies to BSA were elicited in the serum of rabbits by placing 2.6 µg of BSA bound to a strip of nitrocellulose directly in the spleen. To elicit antibodies in the serum of mice required intrasplenic administration of BSA bound to nitrocellulose or Sepharose beads on three or four occasions (Nilsson *et al.*, 1987). Monoclonal antibodies have been produced by giving a single injection of approximately 20 µg of antigen directly into the spleen and performing the spleen cell fusion 3 days later (Spitz & Spitz, 1987).

The technique of immunization of spleen cells *in vitro* is reported to be advantageous for antigens which give a poor immune response or where very small amounts (a few micrograms) of antigen are available (Reading, 1982, 1986; Campbell, 1984; Boss, 1986). It is also a very quick technique as the procedure takes only a few days but this method tends to favour the production of IgM MAbs. However, some refinements to the methods have been published which apparently result in greater numbers of IgG-producing clones (Takahashi *et al.*, 1987).

Cell fusion and hybridoma culture

Appendices 5–12 give details of the methods used for cell fusion and maintenance of fused cells. Other published methods vary in different respects and some of the more important differences and refinements to the methods are discussed below.

Myeloma cell line

There are several suitable cell lines which can be used for mouse or rat fusions (Campbell, 1984; Harlow & Lane, 1988) and they can be obtained from culture collections (American Type Culture Collection, Rockville, MD, USA), commercial companies (ICN Biomedicals Ltd) or other research workers. The main difference is whether or not they retain the capacity to make immunoglobulin. Non-secreting myelomas are usually preferred but there are circumstances where a myeloma secreting a light or heavy chain may be used to advantage (Galfrè & Milstein, 1981; Galfrè & Butcher, 1986). Unless fusions are being done often it is best to keep stocks of myeloma cells frozen and grow up a batch 1–2 weeks prior to the fusion. Myeloma cells grow to approximately $3–6 \times 10^5$ cells ml^{-1} in stationary culture (Kipps & Herzenberg, 1986) and should be in exponential growth (>90% viability) before fusion. The ratio of spleen cells to myeloma cells does not seem to be critical and protocols vary from 1:1 to 10:1 for spleen:myeloma.

It is important to note that myeloma cells contain retroviruses and appropriate precautions should be taken when handling these cell lines (Bartal & Hirshaut, 1987).

Fusogen   Polyethylene glycol (PEG) is very effective in fusing mammalian cells and its properties and possible mode of action have been reviewed by Klebe & Bentley (1987). Polyethylene glycol is toxic to cells and fusion methods depend on bringing 30–50% PEG into contact with cells for a few minutes (Campbell, 1984; Eshhar, 1985). The molecular weight of PEG used by different workers varies from 1500 to 4000. Batches of PEG vary in fusion efficiency and pretested PEG can be obtained from Sigma Chemical Company Ltd and Boehringer Corp. Ltd.

Electrofusion has also been employed to fuse myeloma and spleen cells (Zimmermann, 1987) and some workers claim increased fusion efficiency compared with PEG (Himmler, 1988).

Selection procedures   *HAT vs HAZA*: The fused myeloma and spleen cells, the hybridomas, are selected by using drugs which block the *de novo* synthesis of nucleotides. Spleen cells do not grow well in culture and unfused spleen cells soon die. Myeloma cells are deficient in the enzyme hypoxanthine guanine phosphoribosyl transferase (HPRT) which is utilized in the salvage pathway for making purine nucleotides. Fused cells utilize HPRT from the spleen cell partner so only fused cells should grow in the presence of the drugs (Galfrè & Milstein, 1981). The cell fusion mixture is grown in the presence of hypoxanthine, aminopterin and thymidine (HAT). Aminopterin blocks purine and pyrimidine biosynthesis and the salvage pathways utilize thymidine and hypoxanthine. Alternatively, hybridomas can be selected by inclusion of HAZA (hypoxanthine and azaserine) in the medium. Azaserine blocks purine synthesis and so the medium needs only additional hypoxanthine.

There are a few reasons for preferring HAZA to HAT. Aminopterin is very toxic; it also inhibits glycine biosynthesis and it is metabolized slowly. Care should be taken to wean cells in HT medium for several passages to ensure complete recovery (Eshhar, 1985; Goding, 1986). Also, if the cells are contaminated with mycoplasma, the mycoplasma can cause thymidine deficiency and the fusion will fail (Harlow & Lane, 1988).

AAT selection: The method developed by Taggart & Samloff (1983) uses immunized spleen cells from the RBF/Dn mouse which has, through a chromosome translocation involving chromosomes 12 and 8, linked loci for the heavy chain of immunoglobulin and the selectable enzyme marker adenosine phosphoribosyl transferase (APRT). The spleen cells are fused with FOX−NY myeloma cells, which are APRT−, and the mixtures are grown in medium containing AAT (adenine, aminopterin and thymidine). Antibody-secreting hybridomas are therefore selected. This procedure also allows growth of hybridomas where the spleen cell has lost the X chromosome (and therefore HPRT activity). This method is claimed to produce more viable hybridomas than HAT selection and has proved useful in selecting hybridomas secreting antibodies to rare or weak antigens (Taggart, 1987).

Other forms of hybrid cell selection are reviewed by Shay (1987).

Feeder cells and growth promoters   To aid hybridoma growth some protocols dispense the fusion mixture into culture plate wells containing feeder cells (mouse peritoneal macrophages or spleen cells). Others consider that there are sufficient additional cells from the spleen (monocytes, red blood cells, granulocytes and macrophages) in the fusion mixture so that feeder cells are not necessary (Bartal & Hirshaut, 1987). Some protocols include insulin as an additional additive to the medium or a mixture of oxaloacetate, pyruvate and insulin (Harlow & Lane, 1988) or a growth promoter (Westerwoudt, 1986).

Certain cell lines secrete growth factors into the medium and after active growth for several days and removal of the cells the resultant conditioned medium can be added to cultures to promote hybridoma growth (Campbell, 1984; Westerwoudt, 1986). Conditioned media from mouse cells are available commercially (Sigma; BCL [Boehringer]). Continuously growing cells such as the human cell line MRC-5 (ATCC or ICN Biomedicals Ltd) can support hybridoma growth (Harlow & Lane, 1988), and we have developed a medium supplement which promotes hybridoma growth. It was as good as mouse peritoneal feeder cells for cloning hybridomas and in other instances where cells are growing at low density (Table 1.1; L. Torrance, unpublished results). Furthermore, we found that inclusion of the supplement at fusion gave increased numbers of antibody-secreting hybridomas. We obtained 129/192 wells containing hybridomas secreting specific antibodies when conditioned medium was added to the fusion mixture compared to 17/192 when it was not (L. Torrance, unpublished results). The supplement can be purchased from Life Technologies Ltd.

Inclusion of feeder cells or conditioned medium is essential for successful limiting dilution cloning or other culture conditions where cells are growing at very low density. The use of conditioned medium supplements is preferable to feeder cells as their production can be standardized to give reproducible results, and their use avoids unnecessary sacrifice of animals and the concomitant risk of microbial infection of valuable lines from the feeder cell preparation.

Cell culture   The cells should be fed with medium containing HAZA (100 µg per well) after 4−7 days, and after 7−10 days the plates should be checked for the presence of hybridomas. Mouse hybridoma colonies will be seen under an inverted microscope ($\times 50-100$) as small discrete groups of bright cells. Rat hybridomas tend to be more mobile and do not form discrete clumps. They should be tested for antibody production when the groups are just visible to the naked eye. The cells are fed with medium containing HAZA for the first 2 weeks after the fusion and then with medium containing hypoxanthine for two to three cell generations. Hybridoma CSF should be checked for antibody activity twice and then expanded into 24-well plates when the cells are confluent. Conditioned medium or feeder cells are sometimes necessary at this transfer. The original fusion plates are kept until all the hybridomas of interest have been successfully transferred. It is also worthwhile checking the original

Table 1.1 Comparison of the effect of conditioned medium supplement and mouse peritoneal feeder cells in cloning hybridomas by limiting dilution

	CM	Feeders	No additives
Hybridoma 1	8*	5	0
Hybridoma 2	6	3	0
Hybridoma 3	8	5	0

* Number of wells containing growing hybridomas in a 96-well plate.
CM = conditioned medium.

fusion plates for any late-developing hybridomas. CSF from the microtitre plates can be diluted up to fourfold before testing, thus allowing three or four different assays to be done. It is important to keep a good record of the origin and history in culture of the many different hybridomas. The cell lines can be numbered as follows: LT1/1/52; LT1 is the designation of the fusion; 1 is the number of the original fusion plate (can be 1−5) and 52 is the number of the well in the fusion plate. When cells are cloned the clone can be numbered similarly, i.e. LT1/1/52.239; the point means cloned; 2 is the plate number and 39 is the well number.

Cell cloning   Once wells containing hybridomas secreting the antibody of interest have been identified and the cell lines expanded, the cells should be cloned. If there are more cell lines of interest than can be easily handled at one time (>10), then they can be grown up in 5 ml flat-sided tubes and frozen for regrowth and cloning at a later date (the CSF from frozen lines must be tested for antibody activity). Methods have been published for freezing cells in 96-well tissue culture plates (Wells & Price, 1983; de Liej *et al.*, 1987). However, sometimes the hybridomas of interest are not present in sufficient numbers at such an early stage to survive freezing and it is best to try to clone the most promising lines first.

The usual practice is to clone cells quickly to avoid the problem of the desired cells being outgrown by unstable or non-antibody-producing cells in the culture. However, newly formed hybridomas tend to lose chromosomes until a stable genotype is attained, and they may stop secreting antibody quite quickly. Some workers believe that it is more likely that cell lines stop producing antibody because of loss or inactivation of chromosomes than by overgrowth by non-producing cells, and so they prefer to wait a few weeks before cloning and expending the necessary effort (Westerwoudt, 1986; Candlish, this volume).

We do an initial cloning of cells which are growing in 24-well plates by the quick method (Appendix 9), and then freeze adequate supplies of the cloned cells (Appendix 11); freeze three to four vials on each of three separate occasions. The antibodies are then investigated in more detail and if they seem valuable we clone the line(s) again by limiting dilution (Method 1, Appendix 9). Cloning is usually done until all of the subclones secrete specific antibody. Stock cultures of hybridomas should not be kept growing in culture continuously because they can lose chromosomes over successive cell generations and stop producing antibody, and they may become contaminated. Adequate supplies of vials (at least 12) of cells which have been grown for only a few cell generations after cloning should be kept frozen as stock cultures.

Antibody isotype   It is usually only necessary to determine the type of heavy chain in the first instance, as approximately 95% of mouse (and rat) light chains are κ (Kimball, 1986). There are several different commercial kits available based on immunodiffusion (ICN Biomedicals Ltd; Serotec; Sigma), red blood cell agglutination (Serotec) or ELISA (Zymed Laboratories Inc.), and most are designed to be used with undiluted CSF.

Application of monoclonal antibodies to virus identification

Monoclonal antibodies are not yet widely used in detection assays for plant viruses. This is probably because of the ready availability of good quality polyclonal antisera for several viruses, limited knowledge of the performance of MAbs in routine assays, and limited numbers of suitable MAbs.

That MAbs are epitope specific rather than virus specific confers several advantages. The assay can be very selective because MAbs will differentiate closely related virus strains. For example, the MAV and PAV strains of barley yellow dwarf luteovirus (BYDV) are difficult to distinguish with polyclonal antisera, they are transmitted with different efficiencies by two species of aphid, and infection and spread in the cereal crop depends greatly on the incidence of the vectors (Plumb, 1983). MAbs specific for each strain have been useful in studies on virus incidence and spread (Torrance et al., 1986 b; Pead & Torrance, 1988).

Diagnosis of BYDV strains has to a large extent depended on time-consuming aphid transmission tests. Supplies of strain-specific polyclonal antisera are scarce because BYDV particles are difficult to purify in large quantities free of host plant protein. Use of the MAbs in ELISA has allowed many more strains to be identified accurately (Plumb et al., 1988; Barker, 1990; Burnett & Mezzalama, 1990) and non-specific reactions against host plant proteins to be avoided, thus making the assays more sensitive.

The property of epitope specificity can be further exploited by selecting MAbs which react with epitopes common to a large number of viruses in a group. MAbs against tobacco mosaic virus have been produced which react with other tobamoviruses (Briand et al., 1982), and an anti-potato virus Y MAb has been found to react with 35 other potyviruses tested (Jordan & Hammond, 1988). This latter MAb is offered for sale by Agdia Inc. in a test kit to detect viruses in the potyvirus group.

If MAbs are to be used in routine diagnostic tests, they should ideally possess the following properties: (i) they should react with stable epitopes (parts of the coat protein not susceptible to degradation during sample preparation); (ii) the epitopes should be common to all of the isolates or strains which are to be detected; (iii) the MAbs should be stable on storage; and (iv) they should retain specific activity when bound to microtitre plates or conjugated with enzyme or other molecules. Because it is difficult to exclude the possibility of exceptions to the second property it may be advisable to use an assay incorporating MAbs with two different epitope specificities. It would also be advisable to test several isolates of the target virus or virus group from different geographical locations.

A major advantage in incorporating MAbs into commercial diagnostic kits is the assured continuing supply of high quality reagent. The development of commercial assays requires a large investment of effort in establishing the assay parameters (optimum incubation times, dilution of reagents, test sensitivity, reproducibility, shelf life and quality control procedures). Once these parameters have been established for a MAb-based assay, they should be the same for each batch. However, because of the high cost of development of a panel of MAbs, commercial kit manufacture is probably not economically viable unless the test will be used on a large-scale, such as in monitoring potato stocks entered for certification schemes, or to screen plants in microculture for possible virus infection (Cassells, this volume; Gugerli, this volume).

References

Abu Salih H.S., Murant A.F. & Daft M.J. (1968) The use of antibody-sensitised latex particles to detect plant viruses. *Journal of General Virology* **3**, 299–302.

Adams R.L.P. (1981) Cell Culture for Biochemists. In Work T.S. & Burdon R.H. (eds) *Laboratory techniques in Biochemistry and Molecular Biology* Vol. 8. Elsevier, Amsterdam.

Bahner I., Lamb J., Mayo M.A. & Hay R.T. (1990) Expression of the genome of potato leafroll virus: read-through of the coat protein termination codon *in vivo*. *Journal of General Virology* **71**, 2251–6.

Ball E.M. (1974) *Serological Tests for the Identification of Plant Viruses.* The American Phytopathological Society, St Paul, Minnesota.

Barbara D.J. & Clark M.F. (1986) Immunoassays in plant pathology. In Wang T.L. (ed.) *Immunology in Plant Science*, pp. 197–219. Cambridge University Press Cambridge.

Barker I. (1990) Barley yellow dwarf in Britain. In Burnett P.A. (ed.) *World Perspectives on Barley Yellow Dwarf*, pp. 39–41. CIMMYT, Mexico.

Bartal A.H. & Hirshaut Y. (1987) Current methodologies in hybridoma formation. In Bartal A.H. & Hirshaut Y. (eds) *Methods of Hybridoma Formation*, pp. 1–39. Humana Press, New Jersey.

Bercks R. (1967) Methodische Untersuchungen über den serologischen Nachweis Pflanzen-pathogener Viren mit dem Bentonit-Flockungstest, dem Latex-Test und dem Bariumsulfat-Test. *Phytopathologische Zeitschrift* **58**, 1–17.

Boss B.D. (1986). An improved *in vitro* immunisation procedure for the production of monoclonal antibodies. In Langone J.J. & van Vunakis H. (eds) *Methods in Enzymology* Vol. 121, pp. 27–33. Academic Press, London.

Briand J.P., Al Moudallal Z. & van Regenmortel M.H.V. (1982) Serological differentiation of tobamoviruses by means of monoclonal antibodies. *Journal of Virological Methods* **5**, 293–300.

Burnett P.A. & Mezzalama M. (1990). Barley yellow dwarf virus and aphids in Mexico. In Burnett P.A. (ed.) *World Perspectives on Barley Yellow Dwarf*, pp. 21–4. CIMMYT, Mexico.

Campbell A.M. (1984) Monoclonal antibody technology. In Burdon R.H. & van Knippenberg P.H. (eds) *Laboratory Techniques in Biochemistry and Molecular Biology* Vol. 13. Elsevier, Amsterdam.

Clark M.F. & Adams A.N. (1977) Characteristics of the microplate method of enzyme-linked immunosorbent assay for the detection of plant viruses. *Journal of General Virology* **34**, 475–83.

Cooper J.I. & Edwards M.L. (1986) Variations and limitations of enzyme-amplified immunoassays. In Jones R.A.C. & Torrance L. (eds) *Developments and Applications in Virus Testing*, pp. 139–54. Association of Applied Biologists, Wellesbourne.

de Leij L., Schwander E. & The T.H. (1987) Cryopreservation in hybridoma production. In Bartal A.H. & Hirshaut Y. (eds) *Methods of Hybridoma Formation*, pp. 419–27. Humana Press, New Jersey.

Dietzgen R.G. & Sander E. (1982) Monoclonal antibodies against a plant virus. *Archives of Virology* **74**, 197–204.

Dresser D.W. (1986) Immunization of experimental animals. In Weir D.M., Herzenberg L.A., Blackwell C. & Herzenberg L.A. (eds) *Handbook of Experimental Immunology*, Vol. 1, pp. 8.1–8.21. Blackwell Scientific Publications, Oxford.

Eshhar Z. (1985) Monoclonal antibody strategy and techniques. In Springer T.A. (ed.) *Hybridoma Technology in the Biosciences and Medicine*, pp. 3–41. Plenum Press, London.

Ferguson M., Evans D.M.A., Magrath D.I., Minor P.D., Almond J.W. & Schild G.C. (1985) Induction by synthetic peptides of broadly reactive, type-specific neutralizing antibody to poliovirus type 3. *Virology* **143**, 505–15.

Fox P.C., Berenstein E.H. & Siraganianr P. (1981) Enhancing the frequency of antigen-specific hybridomas. *European Journal of Immunology* **11**, 431–4.

Freshney R.I. (1987) *Culture of Animal Cells. A Manual of Basic Technique*, 2nd edn. Alan R. Liss, New York.

Galfrè G. & Butcher G.W. (1986) Making antibodies. In Wang T.L. (ed.) *Immunology in Plant Science*, pp. 1–25. Cambridge University Press, Cambridge.

Galfrè G. & Milstein C. (1981) Preparation of monoclonal antibodies: strategies and procedures. In Langone J.J. & van Vunakis H. (eds) *Methods in Enzymology* Vol. 73, pp. 3–46. Academic Press, London.

Goding J.W. (1986) *Monoclonal Antibodies: Principles and Practice*, 2nd edn. Academic Press, London.

Gugerli P. & Fries P. (1983) Characterization of monoclonal antibodies to potato virus Y and their use for virus detection. *Journal of General Virology* **64**, 2471–7.

Halk E.L., Hsu H.T., Aebig J. & Franke J. (1984) Production of monoclonal antibodies against three ilarviruses and alfalfa mosaic virus and their use in serotyping. *Phytopathology* **74**, 367–72.

Harlow E. & Lane D. (1988) *Antibodies: A Laboratory Manual*. Cold Spring Harbor Laboratory, Cold Spring Harbor, New York.

Herbert W.J. & Kristensen F. (1986) Laboratory animal techniques for immunology. In Weir D.M., Herzenberg L.A., Blackwell C. & Herzenberg L.A. (eds) *Handbook of Experimental Immunology* Vol. 4, pp. 133.1–133.36. Blackwell Scientific Publications, Oxford.

Himmler G. (1988) Electrofusion vs PEG. In Boonekamp P.M. (ed.) *Monoclonal Antibodies and Immunological Techniques to Detect Plant Pathogens*, p. 17. Pudoc, Wageningen.

Hsu H.T. (1984) Collaborative efforts for producing monoclonal antibodies for the national bank of plant virus antisera at the American type culture collection. In Stern N.J. & Gamble H.R. (eds) *Hybridoma Technology in Agricultural and Veterinary Research*, pp. 232–5. Rowman & Allanheld, New Jersey.

Hudson L. & Hay F.C. (1980) *Practical Immunology*, 2nd edn. Blackwell Scientific Publications, Oxford.

Jordan R. & Hammond J. (1988) Epitope specificity of strain-, virus-, subgroup-specific and potyvirus group cross-reactive monoclonal antibodies. *Phytopathology* **78**, 1600.

Kimball J.W. (1986) *Introduction to Immunology*, 2nd edn. Macmillan, New York.

Kipps T.J. & Herzenberg L.A. (1986) Schemata for the production of monoclonal antibody-producing hybridomas. In Weir D.M., Herzenberg L.A., Blackwell C. & Herzenberg L.A. (eds) *Handbook of Experimental Immunology* Vol. 4, pp. 108.1–108.9. Blackwell Scientific Publications, Oxford.

Klebe R.J. & Bentley K.L. (1987). Chemically mediated cell fusion. In Bartal A.H. & Hirshaut Y. (eds) *Methods of Hybridoma Formation*, pp. 77–96. Humana Press, New Jersey.

Koenig R. & Bercks R. (1968) Anderungen im heterologen Reaktionsvermogen von Antiseren gegen Vertreter der Potato virus X-Gruppe im Laufe des Immunisierungsprozesses. *Phytopathologische Zeitschrift* **61**, 382–98.

Koenig R. & Torrance L. (1986) Antigenic analysis of potato virus X by means of monoclonal antibodies. *Journal of General Virology* **67**, 2145–51.

Mackenzie D.J. & Tremaine J.H. (1986) The use of a monoclonal antibody specific for the N-terminal region of southern bean mosaic virus as a probe of virus structure. *Journal of General Virology* **67**, 727–35.

Makela O. & Seppala I.J.T. (1986) Haptens and carriers. In Weir D.M., Herzenberg L.A., Blackwell C. & Herzenberg L.A. (eds) *Handbook of Experimental Immunology* Vol. 1, pp. 3.1–3.13. Blackwell Scientific Publications, Oxford.

Massalski P.R. & Harrison B.D. (1987) Properties of monoclonal antibodies to potato leafroll luteovirus and their use to distinguish virus isolates differing in aphid transmissibility. *Journal of General Virology* **68**, 1813–21.

Mowat W.P., Dawson S. & Duncan G.H. (1989) Production of antiserum to a non-structural potyviral protein and its use to detect narcissus yellow stripe and other potyviruses. *Journal of Virological Methods* **25**, 199–210.

Nakamura R.M., Voller A. & Bidwell D.E. (1986) Enzyme immunoassays: heterogeneous and homogeneous systems. In Weir D.M., Herzenberg L.A., Blackwell C. & Herzenberg L.A. (eds) *Handbook of Experimental Immunology* Vol. 1, pp. 27.1–27.20. Blackwell Scientific Publications, Oxford.

Nilsson B.O., Svalander P.C. & Larsson A. (1987) Immunization of mice and rabbits by intrasplenic deposition of nanogram quantities of protein attached to Sepharose beads or nitrocellulose paper strips. *Journal of Immunological Methods* **99**, 67–75.

Pead M.T. & Torrance L. (1988) Some characteristics of monoclonal antibodies to a British MAV-like isolate of barley yellow dwarf virus. *Annals of Applied Biology* **113**, 639–44.

Plumb R.T. (1983) Barley yellow dwarf virus — a global problem. In Plumb R.T. & Thresh J.M. (eds) *Plant Virus Epidemiology*, pp. 185–98. Blackwell Scientific Publications, Oxford.

Plumb R.T., Barker I., Forde S. *et al.* (1988) Barley yellow dwarf virus, strain identification. *Rothamsted Annual Report for 1987*, p. 74. Rothamsted Experimental Station, Harpenden, Herts.

Reading C.L. (1982) Theory and methods for immunization in culture and monoclonal antibody production. *Journal of Immunological Methods* **53**, 261–91.

Reading C.L. (1986) *In vitro* immunisation for the production of antigen-specific lymphocyte hybridomas. In Langone J.J. & van Vunakis H. (eds) *Methods in Enzymology* Vol. 121, pp. 18–27. Academic Press, London.

Roberts I.M. (1985) Immunoelectron microscopy of extracts of virus-infected plants. In Harris J.R. & Horne R.W. (eds) *Electron Microscopy of Proteins* Vol. 5, pp. 293–357. Academic Press, New York.

Sela M. (1986) Overview: antigens. In Weir D.M., Herzenberg L.A., Blackwell C. & Herzenberg L.A. (eds) *Handbook of Experimental Immunology* Vol. 1, pp. 1.1–1.7. Blackwell Scientific Publications, Oxford.

Shay J.W. (1987) Mechanism of cell fusion and selection in the generation of hybridomas. In Bartal A.H. & Hirshaut Y. (eds) *Methods of Hybridoma Formation*, pp. 63–75. Humana Press, New Jersey.

Shukla D.D. & Ward C.W. (1989) Structure of potyvirus coat proteins and its application in the taxonomy of the potyvirus group. *Advances in Virus Research* **36**, 273–314.

Shukla D.D., Strike P.M., Tracy S.L., Gough K.H. & Ward C.W. (1988) The N and C termini of the coat proteins of potyviruses are surface-located and the N terminus contains the major virus-specific epitopes. *Journal of General Virology* **69**, 1497–1508.

Spitz M. (1986) 'Single-shot' intrasplenic immunization for the production of monoclonal antibodies. In Langone J.J. & van Vunakis H. (eds) *Methods in Enzymology* Vol. 121, pp. 33–41. Academic Press, London.

Spitz M. & Spitz L. (1987) Intrasplenic immunization for the production of monoclonal antibodies. In Bartal A.H. & Hirshaut Y. (eds) *Methods of Hybridoma Formation*, pp. 249–55. Humana Press, New Jersey.

Stahli C., Staehelin T., Miggiano V., Schmidt J. & Haring P. (1980) High frequencies of antigen-specific hybridomas: dependence on immunization parameters and prediction by spleen cell analysis. *Journal of Immunological Methods* **32**, 297–304.

Taggart R.T. (1987) Culture methods for the selection and isolation of stable antibody-producing murine hybridomas. In Bartal A.H. & Hirshaut Y. (eds) *Methods of Hybridoma Formation*, pp. 181–94. Humana Press, New Jersey.

Taggart R.T. & Samloff I.M. (1983) Stable antibody-producing murine hybridomas. *Science* **219**, 1228–30.

Takahashi M., Fuller S.A. & Hurrell J.G.R. (1987) Production of IgG-producing hybridomas by *in vitro* stimulation of murine spleen cells. *Journal of Immunological Methods* **96**, 247–53.

Torrance L. & Jones R.A.C. (1981) Recent developments in serological methods suited for use in routine testing for plant viruses. *Plant Pathology* **30**, 1–24.

Torrance L., Larkins A.P. & Butcher G.W. (1986a) Characterization of monoclonal antibodies against potato virus X and comparison of serotypes with resistance groups. *Journal of General Virology* **67**, 57–67.

Torrance L., Pead M.T., Larkins A.P. & Butcher G.W. (1986b) Characterization of monoclonal antibodies to a U.K. isolate of barley yellow dwarf virus. *Journal of General Virology* **67**, 549–56.

van Lent J.W.M. & Verduin B.J.M. (1986) Detection of viral protein and particles in thin sections of infected plant tissue using immunogold labelling. In Jones R.A.C. & Torrance L. (eds) *Developments and Applications in Virus Testing*, pp. 193–211. Association of Applied Biologists, Wellesbourne.

van Regenmortel M.H.V. (1982) *Serology and Immunochemistry of Plant Viruses*. Academic Press, London.

van Regenmortel M.H.V. (1986) The potential for using monoclonal antibodies in the detection of plant viruses. In Jones R.A.C. & Torrance L. (eds) *Developments and Applications in Virus Testing*, pp. 89–101. Association of Applied Biologists, Wellesbourne.

van Regenmortel M.H.V. (1989) The concept and operational definition of protein epitopes. *Philosophical Transactions of the Royal Society of London*, **B 323**, 451–66.

van Regenmortel, M.H.V., Briand J.P., Muller S. & Plaue S. (1988) Synthetic polypeptides as antigens. In Burdon R.H. & van Knippenberg P.H. (eds) *Laboratory Techniques in Biochemistry and Molecular Biology*. Elsevier, Amsterdam.

van Regenmortel M.H.V. & Neurath A.R. (1985) Structure of viral antigens. In van Regenmortel M.H.V. & Neurath A.R. (eds) *Immunochemistry of viruses. The Basis for Serodiagnosis and Vaccines*, pp. 1–11. Elsevier, Amsterdam.

Voller A., Bartlett A., Bidwell D.E., Clark M.F. & Adams A.N. (1976) The detection if viruses by enzyme-linked immunosorbent assay (ELISA). *Journal of General Virology* **33**, 165–67.

Weir D.M., Herzenberg L.A., Blackwell C. & Herzenberg L.A. (1986) *Handbook of Experimental Immunology*, 4th edn, Vol. 1. Blackwell Scientific Publications, Oxford.

Wells D.E. & Price P.J. (1983) Simple rapid methods for freezing hybridomas in 96-well microculture plates. *Journal of Immunological Methods* **59**, 49–52.

Westerwoudt R.J. (1986) Factors affecting production of monoclonal antibodies. In Langone J.J. & van Vunakis H. (eds) *Methods in Enzymology*, Vol. 121, pp. 3–18. Academic Press, London.

Zimmerman U. (1987) Electrofusion of cells. In Bartal A.H. & Hirshaut Y. (eds) *Methods of Hybridoma Formation*, pp. 97. Humana Press, New Jersey.

Appendix 1: Preparation of antigen in Freund's adjuvant

Freund's adjuvant is composed of a mixture of mineral oils and can be obtained from Sigma or Pierce and Warriner Ltd (for addresses of suppliers see p. 230). The complete Freund's adjuvant (CFA) contains heat-killed *Mycobacterium tuberculosis*; the incomplete does not contain bacteria. Usually an equal volume of CFA is emulsified with antigen (diluted in weak neutral phosphate buffer or saline). Use glass Luer lock syringes connected together by a syringe adapter (Aldrich) or three-way valve and pass the aqueous solution into the CFA first, then squeeze the mixture back and forth between the syringes. Test the emulsion by putting a small drop onto water; if the drop stays intact the emulsion is ready to use, if the drop disperses then continue mixing and if necessary add more CFA. Glass syringes with Luer locks are essential for work with CFA because the emulsion becomes very stiff and other connectors may come apart spraying adjuvant over the operator. Never inject emulsions of adjuvant intravenously. The syringes can be cleaned with acetone before washing with detergent. Care should be taken when handling Freund's adjuvant as it can cause eye damage, severe hypersensitive reactions, or permanently stiff fingers if injected accidentally (Goding, 1986).

Details of how to make different types of injections are given by Herbert & Kristensen (1986). However, in the UK all such procedures must be performed by trained personnel under appropriate authority from the Home Office.

Appendix 2: ELISA

There are many forms of ELISA, the format chosen depending on the particular application. Two formats often used in plant pathology are:

• An indirect format where antigen is trapped onto microtitre plates by a layer of homologous polyclonal antibodies (antibody trapped antigen, ATA).

• An indirect format where antigen is adsorbed directly to the plate (plate trapped antigen, PTA).

These methods are well suited for the initial screening of MAb cell lines raised to viruses or their antigens, although it may be necessary to use other types of format if a second round of screening is to be done. The same formats are often used in the rapid, large-scale detection of viruses in plant material.

The materials and methods for several forms of ELISA are described below.

Materials
Buffers

• Phosphate buffered saline (PBS), pH 7.4 ($5\times$ conc.)

NaCl	40 g
KH_2PO_4	1 g
$Na_2HPO_4.12H_2O$	14.5 g
KCl	1 g

Make up to 1 litre and store at 4°C.

• Washing buffer, PBST

$5\times$ conc. PBS	200 ml
10% Tween 20	5 ml

Make up to 1 litre.

• Coating buffer, pH 9.6

Na_2CO_3	1.59 g
$NaHCO_3$	2.93 g

Make up to 1 litre.

• Substrate buffer for alkaline phosphatase

Diethanolamine	97 ml
Distilled water	800 ml

Adjust pH to 9.8 with HCl.
Make up to 1 litre.
Add 0.5 mg nitrophenyl phosphate per 10 ml.
Stop reaction with 3 M NaOH (optional). Read absorbance at 405 nm.

• Substrate buffer for horseradish peroxidase
3,3',5,5'-tetramethylbenzidine (TMB) and hydrogen peroxide.
Stock solutions:
10 mg ml^{-1} TMB
6% hydrogen peroxide
1 M sodium acetate/citric acid buffer, pH 6.
For each plate mix (just before use) 9 ml H_2O, 1 ml buffer, 0.1 ml TMB, 10 µl H_2O_2.
Stop with 25 µl H_2SO_4 and read absorbance at 450 nm.

Blocking solutions

Various solutions can be used; the four most common are:
10% newborn calf or foetal calf serum (FCS)
10% horse serum

0.1−10% non-fat dried milk
Tween 20.

Enzyme conjugates and other reagents   Good quality anti-species, e.g. anti-mouse or anti-rabbit immuno-globulin, alkaline phosphatase, horseradish peroxidase, protein A, biotin, etc. conjugates are available from several suppliers (Sigma, Bio-Rad, BCL (Boehringer), Pierce, Cambridge BioScience (Zymed), Serotec and ICN Biomedicals Ltd).

Methods
Direct double antibody sandwich ELISA
- Dilute antibody stock solution (1 mg ml^{-1}) in coating buffer (the correct dilution must be determined by experiment, but it is usually 1/500 or 1/1000). Use a multichannel pipette to dispense the diluted antibody into the wells of a microtitre plate, 100 µl per well, cover and incubate for 2−3 h at 37°C.
- Rinse the plate three times with PBST by successively filling and emptying the wells.
- Add virus-infected or control samples to the wells, cover and incubate the plate in the refrigerator overnight.
- Rinse the plate three or four times with PBST.
- Add alkaline phosphatase conjugated virus-specific antibodies diluted in PBST containing 0.2% BSA. Cover and incubate for 2 h at 37°C.
- Rinse plate three times with PBST.
- Add freshly prepared substrate, cover and leave at room temperature for colour to develop.

Indirect ELISA
PTA (plate trapped antigen)
- Dilute virus preparation in coating buffer or PBS and pipette 100 µl of solution into the wells of a microtitre plate, cover and leave in the refrigerator overnight. (This can be done in advance using buffer containing sodium azide and left in the refrigerator until needed.)
- Rinse plates three times with PBST.
- Include a blocking step if appropriate by adding 200 µl of blocking solution per well for 1 h at room temperature.
- Add virus-specific antibody or hybridoma culture supernatant fluid (dilute with PBST + BSA). Cover and incubate at room temperature or 37°C for 2 h.
- Rinse plate three times with PBST.
- Add enzyme-conjugated anti-species antibody diluted (usually 1/1000) with PBST + BSA. Cover and incubate for 2 h at 37°C.
- Rinse as above.
- Add freshly prepared substrate solution, cover and leave at room temperature for colour to develop.

ATA (antibody trapped antigen)
- Coat microtitre plate with virus-specific antibody by diluting antibodies in coating buffer and adding 100 µl of solution to each well. Incubate for 2 h at 37°C.
- Rinse plate three times with PBST.

- Add virus-containing and control samples, incubate overnight in a refrigerator.
- Rinse plate three or four times with PBST.
- Add second antibody, e.g. CSF. Incubate for 2 h at room temperature or 37°C.
- Rinse plate three times with PBST.
- Add enzyme-conjugated anti-species antibody diluted (usually 1/1000) with PBST + BSA. Cover and incubate for 2 h at 37°C.
- Rinse plate three times with PBST.
- Add freshly prepared substrate solution and leave at room temperature for colour to develop.

Notes on experimental technique

- All reagents are used at 100 μl per well (except block solutions which are used at 200 μl per well).
- Antibodies or enzyme conjugates are usually used at dilutions of 1/500–1/2000, and CSF at 1/4–1/100, but the correct dilution must be checked by experiment.
- All of the stages of the assay should be done quickly so that the wells of the plate do not dry out.
- Plant sap should be extracted by crushing the leaves in a mortar and pestle and filtering the pulverized leaves through muslin.
- The plate washing procedure must be thorough, ensuring that all wells are completely filled and emptied at each rinse and that rinsing is sufficient to remove all of the green sap. For sap a good method of rinsing is to empty the wells, then fill with PBST, shake out the buffer immediately and refill the wells, then leave to soak for 2–3 min. Repeat the rinse and soak procedure a further two or three times. For other reagents it is sufficient to give three rinses without soaking.
- Microtitre plates can be of variable quality; for plant virus work Nunc Immunoplate II (Life Technologies Ltd) or Linbro (ICN Biomedicals Ltd) give good results.

Appendix 3: Assay of cell culture supernatant fluids for immunoglobulin by dot blot on nitrocellulose membrane

Materials
- Nitrocellulose membrane (Schleicher & Schuell, 0.45 μm).
- Tris buffered saline (TBS)
    Tris   6 g
    NaCl   8 g
    Adjust pH to 7.4 with HCl and dilute to 1 litre.
- TBS–Tween: TBS containing 0.1% Tween 20.
- Dried skimmed milk powder.
- Anti-mouse alkaline phosphatase conjugate.
- NBT (*p*-nitro blue tetrazolium chloride) 30 mg ml^{-1} in 70% dimethyl formamide.
- BCIP (5-bromo-4-chloro-3-indolyl phosphate *p*-toluidine salt) 15 mg ml^{-1} in dimethyl formamide.
- Substrate buffer: 12.1 g Tris + 5.8 g NaCl + 1 g MgCl$_2$.
Adjust pH to 9.5 with HCl and dilute to 1 litre.

Method
- Wet the nitrocellulose with distilled water, then place in BIO-DOT^{TM} apparatus (e.g. Bio-Rad).

• Add 50−100 µl CSF supernatant diluted in TBS. Suck through the membrane.
• Remove membrane from apparatus and incubate in blocking solution (TBS containing 0.1% Tween 20 and 5% skimmed milk powder) for 1 h at room temperature.
• Add anti-mouse alkaline phosphatase conjugate (usually diluted 1/1000) in blocking solution for 1 h at room temperature.
• Wash three times for 5 min in TBS−Tween.
• Add NBT/BCIP, 100 µl of each to 10 ml substrate buffer.
• Stop reaction after approximately 15 min by rinsing in water.

Appendix 4: Immunoblotting technique

Materials Use the same materials as in Appendix 3. The proteins are transferred from a polyacrylamide gel to the nitrocellulose membrane electrophoretically in a Trans-Blot apparatus (Bio-Rad; refer to manufacturer's instructions).

Method • After blotting the membrane is placed in blocking solution (TBS containing 0.1% Tween 20 and 5% skimmed milk) for 1 h at room temperature.
• Wash the membrane three times in baths of TBS−Tween with gentle shaking.
• Incubate the membrane with MAb appropriately diluted in TBS-Tween containing 0.5% BSA in sealed plastic bags or a multichannel apparatus for 2 h at room temperature.
• Wash in baths of TBS−Tween as above.
• Incubate with anti-mouse (or anti-rat) alkaline phosphatase conjugate diluted in TBS−Tween−BSA for 2 h at room temperature.
• Wash as above.
• Add NBT/BCIP, 100 µl of each to 10 ml substrate buffer.
• Stop the reaction by rinsing the membrane in water.

Appendix 5: Media used for cell fusions and maintenance of cell lines

Media and additives can be obtained from Life Technologies Ltd, ICN Biomedicals Ltd or Sigma Chemical Company Ltd.

Media Temperatures at which media should be stored are indicated.

• RPMI 1640 (1× conc.) without glutamine	4°C
• FCS Myoclone Plus (Life Technologies Ltd)	−70°C
• L-glutamine (100× conc.) liquid (G)	−20°C
• Hypoxanthine (100× conc.) liquid	4°C
• Penicillin−streptomycin solution (PS, 10 000 units penicillin; 10 000 µg streptomycin)	−20°C
• Gentamicin liquid (50 mg ml^{-1})	4°C
• Amphotericin B (F, fungizone liquid 250 µg ml^{-1})	−20°C
• PEG (Sigma, 5 g mol. wt 3000−3700)	Room temp.
• HAZA (50× conc.) (Sigma)	−20°C
• Dimethyl sulphoxide, DMSO (Sigma)	Room temp.
• Conditioned medium supplement (CM) (Life Technologies Ltd)	−20°C

Make up medium as follows:

  100 ml RPMI + 1 ml G + FCS.

For 20% FCS add 25 ml; 15% FCS add 18 ml; 10% FCS add 12 ml; and 7% FCS add 8 ml. Serum concentrations are approximate. It is important to reduce the amount of serum used as quickly as possible to reduce unnecessary expense. At fusion and for the first feeds use 15–20% FCS but reduce to 10–15% FCS for subsequent culture. Reduce to 8–10% FCS for established lines; some lines grow well in only 2.5% FCS. Established hybridomas can grow in serum-free media or media with mixtures of additives designed to reduce the serum levels considerably such as Nutridoma™ (BCL), CPSR™ (Sigma), Ultroser™, OptiMEM™ and serum-free hybridoma medium (Life Technologies Ltd). To reduce expense newborn calf serum or horse serum can also be used but they may contain immunoglobulins which cause non-specific reactions in assays.

Penicillin–streptomycin (PS) and fungizone (F) are added at 1 ml per 100 ml RPMI at the fusion and when culturing cells after storage in liquid nitrogen but should not be used routinely. Routine use of antibiotics risks the development of resistant strains and conceals bad techniques. Gentamicin should be kept in reserve for use when PS does not control the problem.

HAZA is supplied as a 50× concentrate. Dissolve in 10 ml RPMI (injected through the septum of the container), add 5 ml to 250 ml 20% FCS RPMI, and store the remainder in the original vial at −20°C.

PEG is supplied as a sterile waxy solid. Heat, then add aseptically 5 ml of RPMI (serum-free). Store at room temperature with the bottle wrapped in aluminium foil.

Appendix 6: Materials to have ready for fusion

Media
- 200 ml RPMI + 20% FCS + G + PSF + HAZA + 10% CM (warm)
- 50 ml RPMI + 2.5% FCS + PSF + G
- 100 ml RPMI + PSF + G

Instruments and materials   The following items will be required: forceps and scissors, pipette tips, 70% alcohol, two 10 ml syringes, two 21G needles, 9 cm Petri dishes, 50 ml centrifuge tubes, six microtitre plates, unplugged pipettes, media reservoirs, supplies of universal bottles, 10 ml plugged pipettes, 1 ml plugged pipettes and a multichannel automatic pipette.

Appendix 7: Preparation of spleen and myeloma cells

Spleen   Two pairs of scissors and forceps are required. They should be clean and dipped in alcohol before use. The mouse should be killed (save the blood for future tests), soaked in alcohol and placed on its back for dissection. The first set of scissors and forceps are used to cut and pull back the skin from the abdomen. The second set are used to make a cut in the peritoneal wall on the left side. The spleen is on the left of the peritoneal cavity just below the stomach. The veins and connective tissue by which it is attached all lie on the side of the organ toward the peritoneal cavity.

Collect the spleen in a universal bottle containing 10 ml RPMI 2.5%

FCS + PSF + G (keep on ice until use). Transfer the spleen to a second universal bottle containing fresh medium and wash; repeat two to three times. Transfer the spleen to a Petri dish and remove any fatty or connective tissue, transfer to a second Petri dish containing 5 ml RPMI + 2.5% FCS and then gently tease it apart using 21G needles. Draw the cell suspension into a 10 ml syringe through the 21G needle and pipette out again; repeat three to four times and then pipette into a universal bottle and dilute to 10 ml. Centrifuge for 8 min at 400 g.

Aspirate supernatant fluid and disturb the pellet by flicking the side of the bottle, add 10 ml of serum-free medium and resuspend cells gently. Repeat centrifugation and resuspend cells in 10 ml serum-free medium.

Myeloma cells   Harvest cells, approximately 5×10^5 cells ml^{-1}, >90% viability. Centrifuge at 400 g for 8 min, resuspend cells in 10 ml serum-free medium and count them (count only well-rounded whole cells).

Controls   Save 0.5 ml of each cell type, centrifuge and resuspend in 2.5 ml medium containing HAZA. Pipette 100 µl into each well of one row of a microtitre plate.

Appendix 8: Cell fusion

Add spleen cells to myeloma cells (ratio approximately 5:1, e.g. 1×10^8 mouse spleen cells plus 2×10^7 myeloma cells). Spin for 8 min at 400 g. Aspirate supernatant fluid, dry well (remove all drops of medium) and flick pellet.

While stirring continuously, slowly add with a 1 ml pipette 0.8 ml of PEG (warm), and continue stirring for 1.5 min (clumps of cells should be visible at this stage).

Continue stirring and add warm serum-free medium to the cells at the following rates:

> 1 ml over 1 min;
> 1 ml over 1 min;
> 1 ml over 30 s;
> 1 ml over 30 s;
> 6 ml over 1.5 min.

Then, with a 10 ml pipette, add 15 ml directly with continuous stirring. Leave cells for 5 min. at 37°C.

Centrifuge for 8 min at 400 g. Aspirate supernatant fluid, flick pellet, and add 50 ml RPMI + 20% FCS + HAZA + 10% CM. Plate out the cells, 0.1 ml per well of microtitre plates.

After fusion   Leave cells undisturbed for 5–7 days and then check for the presence of hybridomas. Feed cells with 50–100 µl medium containing HAZA after 7–10 days. Check control cells and record wells which contain growing hybrids.

Appendix 9: Cloning cells

Cell lines should be growing healthily in a 24-well plate, and should have been screened recently for antibody activity. Cloning should be done with 15% FCS RPMI containing 10% CM.

Alternatively, the day before cloning, prepare feeder layers of mouse peritoneal cells, 100 μl per well in 96-well microtitre plates. Prepare four to six plates per cell line to be cloned.

Method 1 Disturb cells of the line to be cloned by sucking the medium into and out of a plastic Pasteur pipette a few times and then transfer a small volume of this to a fresh well. Count the cells and dilute to 50 and 5 cells ml^{-1}. Using a multichannel pipette dispense 100 μl per well into microtitre plates: one to two plates at five cells per well and two to four plates at half a cell per well. Leave undisturbed for 7−10 days and then examine for growth. Screen wells when hybridomas cover about a quarter of the well surface (if possible choose wells which contain only one colony).

Method 2 (quick method) To a microtitre plate containing 100 μl of feeder cells, or 100 μl of medium containing 10% CM, add 100 μl of cell suspension to well number A1. Make one in two doubling dilutions down row one. Using a multichannel pipette make one in two doubling dilutions across the plate. Leave for 7−10 days and check for growth as above.

Note that cloning method 1 (half a cell per well) is more likely to ensure that the colonies have been derived from a single cell. However, there is still the possibility that the cells were clumped together so cell lines should be cloned more than once. With method 1 it is essential to use four to six plates because at such high dilution few hybrids will grow per plate whereas with method 2, one or two plates are usually sufficient.

Appendix 10: Feeder cells

Mouse peritoneal cells Kill the mouse (dislocate its neck), then lay the animal on its back and soak the fur with alcohol, make an incision in the skin and tear it back to expose the peritoneum. Inject 4 ml serum-free medium containing heparin (250 μl per 25 ml RPMI + PS). Agitate the body and withdraw the fluid, transferring it to a universal bottle on ice. Spin at 400 g for 8 min, and resuspend the cells in 40 ml 15% FCS RPMI + PS. Plate out at 0.1 ml per well over four microtitre plates. Cells from one mouse can be used for four to five microtitre plates (approx. 4 × 10^6 cells can be obtained from one mouse and used at approx. 10^4 cells per well).

Spleen cells Kill the mouse as detailed above and remove the spleen, wash twice with serum-free RPMI and remove surplus connective or fatty tissue. Tease the spleen apart with 21G needles and suck up the suspension into a 10 ml syringe three to four times. Transfer to a universal bottle (leaving the larger clumps behind) and centrifuge at 400 g for 8 min. Resuspend in 100 ml 15% FCS RPMI and add 0.1 ml per well to a microtitre plate (leave in incubator for 1 day and check for contamination before use).

Rescuing cell lines

To try to rescue a cell line which is not growing very well, collect spleen cells as above in 50 ml medium containing 20% FCS, and leave the spleen cells in the incubator for 1−2 days to check for contamination. Remove most of the medium from the poorly growing cell line and replace with the spleen cell suspension.

Appendix 11: Freezing cells

Established cell line

Take 10−15 ml of a cell line which is growing well (not overgrown or dying). Centrifuge at 400 g for 8 min, remove the supernatant and save this for assay. Resuspend the pellet in 1 ml freezing medium (90% FCS and 10% DMSO) and dispense 0.5 ml into each of two cryotubes. Put the tubes in a polystyrene box (keep upright) and place in −70°C freezer for 1 day before storing the cryotubes in liquid nitrogen.

Cell lines growing in 24-well plates or 5 ml tubes

Centrifuge 1−5 ml of cell suspension as above, resuspend the pellets in 0.5 ml freezing medium and place in one cryotube.

Note that the cells should be kept cold during the above manipulations and frozen quickly. The tubes of cells are placed in a bucket of wet ice or a cooled metal rack during the manipulations. Use cold freezing medium. The minimum number of cells needed is about $10^6 - 10^7$ per vial for successful recovery. The cells should be frozen at a controlled rate of about 1°C min^{-1} and cooling between 0 and −30°C is critical. The cells can be placed in a polystyrene box in a −70°C freezer overnight, or in a freezing tray (Taylor-Wharton, Jencons (Scientific) Ltd) which sits in the neck of a liquid nitrogen Dewar vessel, and which can be adjusted to lower the cells through different temperature zones. Label the cryotubes with the number of the cell line, thawing medium and date. Fill in a record book (or computer log) for location in liquid nitrogen.

Appendix 12: Thawing cells

Cells should be thawed quickly. Take the vials from the liquid nitrogen store and place them in a water bath (37°C; do not put the seal under water). When the cells have almost thawed dry the freezing vial and swab with alcohol, transfer the cells to a universal bottle and slowly add, with stirring, 10 ml of cold medium (the same composition as the cells were growing in when frozen). Spin at 400 g for 8−10 min, suck off the supernatant fluid and resuspend the cells in 1 ml fresh medium. Plate out on a 24-well plate at one or two dilutions, e.g. 0.5, 0.25, 0.1 ml per well (and add a little more medium where necessary). If there are few cells or there is some doubt about cell viability plate out with medium containing conditioned medium supplement or feeder cells.

Note that care should be taken when removing vials of cells from liquid nitrogen; wear gloves and a face mask to guard against the possibility that the vial may explode.

2 Immunodiagnostic techniques for plant mycoplasma-like organisms

M.F. CLARK

Introduction

In the two decades since the first reports of plant infection by mycoplasma-like organisms (MLO) (Doi *et al.*, 1967) the number of diseases attributed to MLO infection has risen to more than 300 (McCoy *et al.*, 1987). During this period numerous attempts have been made to isolate and culture these wall-less pro-karyotes *in vitro* without success except for the taxonomically more advanced spiroplasmas. The term 'mycoplasma-like' derives largely from ultrastructural studies of the organisms in infected tissue and reflects the morphological and structural similarity of these intracellular prokaryotes with the cul-turable, extracellular animal mycoplasmas.

Attribution of plant disease to an MLO aetiology has been based largely on symptom expression, association with arthropod vectors (particularly leafhoppers) and electron microscopy of thin sections of phloem tissue by which pleomorphic structures with a trilaminar unit membrane may be identified. In some cases symptom remission following chemotherapy with tetra-cyclines has afforded additional evidence of an MLO aetiology (Ishiie *et al.*, 1967; McCoy *et al.*, 1987). Electron microscopy alone is an unreliable diagnostic method because the organisms are unevenly distributed and present in low concentration in the tissue, making sample selection difficult. The use of light microscopy in conjunction with various staining techniques, principally the DNA fluorochromes Hoechst 33258 and 4',6-diamidino-2-phenylindole (DAPI) (Seemüller, 1976) has been more successful.

Although adequate for detecting MLOs and correlating their presence in phloem tissue with the expression of disease symp-toms, these methods are not able to identify or discriminate among different MLOs. The intractable nature of these organ-isms has hampered the production of specific diagnostic re-agents, but recent improvements in techniques have encouraged workers in several laboratories to prepare polyclonal antisera and monoclonal antibodies (MAbs) to selected MLOs from infected plant tissue or vector insects (Sinha, 1979; Caudwell *et al.*, 1982; Clark *et al.*, 1983; Sinha & Benhamou, 1983; Sinha & Chiykowski, 1984; Clark *et al.*, 1986; Lin & Chen, 1986; Hobbs *et al.*, 1987; Clark *et al.*, 1988; Clark *et al.*, 1989). In addition there is a growing interest in the isolation and cloning of MLO genomic DNA and the development of nucleic acid hybridization probes (Kirkpatrick *et al.*, 1987; Davis *et al.*, 1988; E. Seemüller, pers. comm.).

The inability to culture MLOs *in vitro* restricts the source material for the production of diagnostic reagents to infected plants or arthropod vectors. Various procedures of differing complexity for the extraction and partial purification of MLO immunogens from such tissue, preparatory to raising polyclonal antisera or MAbs, have been described (Sinha, 1979; Caudwell *et al.*, 1982; Clark *et al.*, 1983, 1989; Jiang & Chen, 1987). Such immunogen preparations are almost inevitably contaminated to some degree with plant or insect components, although the level to which such contaminants interfere with subsequent immunoassays varies widely among the various reports. This chapter reviews results and experiences from my laboratory in preparing MLO immunogens from plants and in developing various immunoassay procedures suitable for the detection and identification of MLOs in plant tissue, tissue extracts and partially purified preparations.

Preparation of immunogen

The quality of the immunogen preparation is the most crucial aspect in the production of suitable polyclonal antisera, and perhaps to a lesser extent, MAbs. For this reason it is worth considering various aspects of immunogen purification. Selection of the source tissue is the most critical factor. Host species, age and stage of infection should all be such as to maximize the yield of MLOs from a given amount of tissue. In this regard the speed and simplicity of the DAPI fluorescent stain procedure (Seemüller, 1976) (Appendix 4) can be advantageous in monitoring the quantity of MLO in the source tissue. Unfortunately as this procedure is only really applicable to the detection of MLO in tissue sections it is of little use in monitoring the progress of immunogen preparation.

Choice of purification protocol is less easy to make but should reflect the objectives of the study to be undertaken since it is unlikely to prove possible to achieve maximum purity and, at the same time, preserve structural and biological integrity. If the primary requirement is to prepare a diagnostic reagent enabling MLO detection and identification it may be more prudent to forego retention of structural integrity and biological activity in order to maximize antigenic or immunogenic competence. On the other hand if the objective is an investigation of MLO fine structure or of host—pathogen relations preservation of MLO structure and infectivity will be a prime consideration.

In the protocol given in Appendix 1, preservation of morphology is subordinated to retention of serological activity. MLO antigens appear to deteriorate rapidly both with time and with the number of steps in the procedure. This basic protocol is relatively short and should be completed in 1 day. It has been used, with minor modifications, to successfully prepare polyclonal antibodies to more than a dozen MLOs, as well as MAbs to three of them.

Where a suitable antibody is already available the progress of the purification may be monitored by assaying recovery of antigen using an enzyme-linked immunosorbent assay (ELISA) procedure (see below).

The recovery of tomato big bud MLO-associated antigen measured by indirect ELISA at different stages during purification is shown in Fig. 2.1. $F(ab')_2$ fragments of rabbit immunoglobulin purified from antiserum raised against earlier tomato big bud MLO preparations were adsorbed to the wells in a polystyrene microtitre ELISA plate. Samples of the MLO preparations taken at each stage of the procedure were diluted in phosphate buffered saline (PBS) containing $0.5\,\mathrm{ml\,l^{-1}}$ Tween 20 (PBS—Tween) for testing. To enable direct comparisons to be made the first dilution of the samples was equivalent to a $1:20\,(\mathrm{g\,ml^{-1}})$ extract of the original tissue. Following incubation with cross-absorbed IgG (Appendix 2) the presence of MLO antigen was revealed with a protein A—horseradish peroxidase conjugate (reactive only with the Fc portion of the IgG).

Preparation of antibodies

Both polyclonal and monoclonal antibodies may be obtained following immunization with preparations made using the protocol in Appendix 1.

Polyclonal antibodies   Rabbits are immunized with an emulsified 1:1 mixture of the MLO preparation and Freund's adjuvant. Complete adjuvant is used for the primary immunizing injection and incomplete adjuvant for all subsequent injections. About half the emulsion is administered using an intramuscular injection into the thigh, the remainder is injected subcutaneously and intradermally at five to eight sites along the back. Booster injections are given every 3—4 weeks. A test bleed may be taken from the marginal ear vein 1 week after the fourth boost. If the antibody activity is satisfactory additional bleeds can be made at intervals of

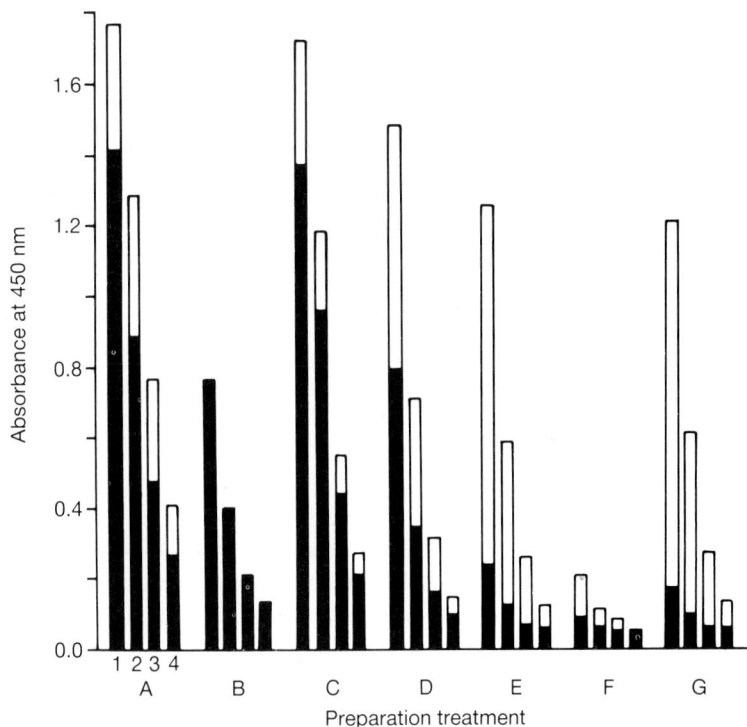

Fig. 2.1 Recovery of ELISA-detectable tomato big bud MLO-associated antigen during purification. MLO-infected (open bars) and healthy (solid bars) plant tissue extracts were treated as described in Appendix 1. Treatments were: A, clarified extract; B, PEG supernatant; C, PEG precipitate; D, cross-absorbed with anti-plant rabbit antiserum; E, after passage through protein A-Sepharose column; F, high-speed sedimentation supernatant; and G, high-speed sedimentation pellet. 1, 2, 3, 4 = antigen preparations diluted 1:20, 1:63, 1:200 and 1:630 respectively, relative to the fresh weight of plant tissue extracted. (From Clark *et al.*, 1989.)

approximately 2 weeks. Further booster injections should be given as necessary to maintain antibody titre.

Antisera obtained by immunizing animals with preparations made as in Appendix 1 will exhibit activity for some host constituents. The level of activity will vary among antibody preparations. For some diagnostic procedures such activity may be sufficiently low as not to interfere unduly with the MLO assay. More often it has been found necessary to reduce anti-host activity by cross-absorbing the antiserum with a preparation of host plant tissue (see Appendix 2).

Monoclonal antibodies   BALB/c mice are immunized using MLO preparations and an injection schedule similar to that used for rabbits, except that the injections are made intraperitoneally, using a maximum of

0.5 ml of the emulsified MLO preparation. The pre-fusion boost is also given intraperitoneally but without adjuvant.

Following fusion of spleen cells with an appropriate mouse myeloma cell line (e.g. NS-1, P3 X63-Ag8.653, NSO/1, or Sp2) (see Torrance, this volume), cells to be cloned are identified by screening cell culture supernatant fluids for specific antibody activity using an indirect method of microplate ELISA. Cell culture supernatant fluids should be tested against both infected and healthy antigen preparations. To reduce the likelihood of anti-MLO antibody activity being masked by possibly more prevalent anti-plant antibodies, the cell culture supernatant fluids should be diluted prior to screening with an equal volume of a 1:40 (g ml^{-1}) extract of healthy plant tissue in PBS containing 1 g l^{-1} Tween 20.

MLO antigens to be used for antibody screening are immobilized on the solid phase either by binding to previously adsorbed specific antibody (capture antibody), or by direct adsorption to the microplate. Antigen preparations suitable for direct adsorption may be obtained by following steps 1–8 of Appendix 1, modified only in that the Sepharose column should be pre-equilibrated and samples eluted with PBS instead of GM buffer. PBS is a more suitable buffer for coating the MLO antigens than either GM or carbonate buffers. Shorter but less effective antigen preparation procedures can also be used, for example using steps 1–5 of Appendix 1, or more simply using steps 1–3 and then step 7. A comparison of the effectiveness of these antigen preparation methods is shown in Fig. 2.2.

Detection and identification procedures

Electron microscopy using gold-labelled antibodies
Electron microscopy of fixed and embedded tissue sections is the conventional method of visually identifying the presence of MLOs in infected plants. This method is satisfactory where the distribution of MLO in the tissue is fairly uniform and the concentration is high, but the main disadvantage is the long time (about 1 week) required to process the tissue sections to the stage of examination. Electron microscopy alone is unable to identify specific MLOs or to discriminate between different MLOs. MLO identification can be achieved using gold labelling of specific antibody (preferably MAb) used to decorate the MLO either in tissue sections or in agarose-embedded pellets obtained from partially purified preparations. A requirement of this technique is that the antibody used is directed to a surface antigen on the MLO. When applying immunogold labelling to fixed tissue pieces prior to embedding and sectioning (see Appendix 3) allowance should be made for the limited extent of pen-

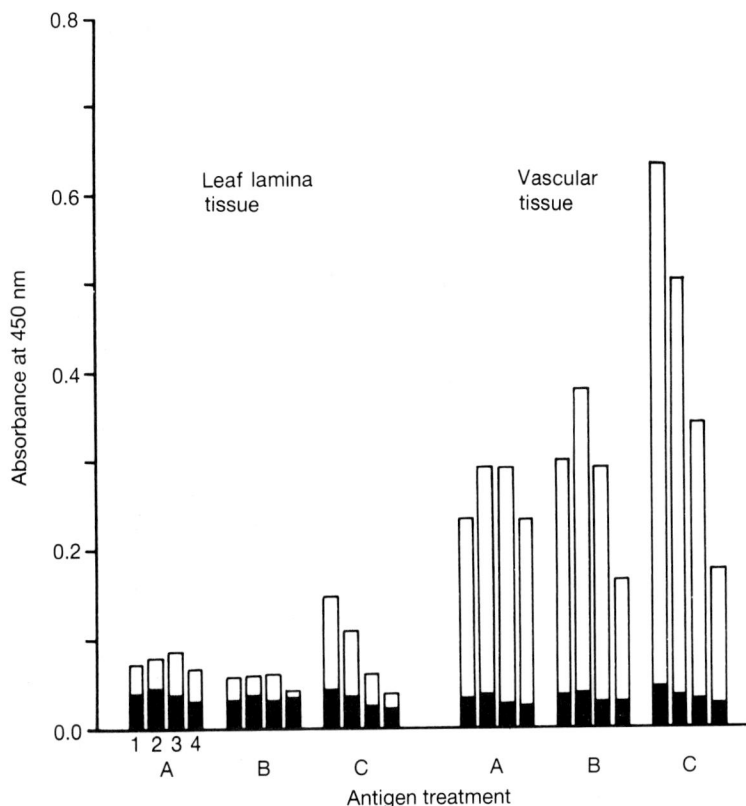

Fig. 2.2 Influence of source and treatment of primula yellows MLO-infected plant tissue on antigen yield and on suitability for directly coating microwell plates for ELISA. Open bars = infected plant extract; solid bars = healthy plant extract. Treatments were: A, antigens after passage through Sepharose 4B; B, antigens precipitated with 60 g l^{-1} PEG 6000; C, PEG-precipitated antigens after passage through Sepharose 4B. 1, 2, 3, 4 = antigen preparations diluted 1:25, 1:80, 1:250 and 1:800 respectively, relative to the fresh weight of plant tissue extracted. Antigens detected with EM1 monoclonal antibody to primula yellows MLO. (From Clark *et al.*, 1989.)

etration of the antibody into the cells from the cut end of the tissue section. For such tests it is advisable to begin sectioning as close to the cut end of the tissue as is practicable.

Immunogold labelling (see Appendix 3) has also been applied with some success to MLO membrane antigens adsorbed directly to the electron microscope grids or trapped by immunosorbent procedures (R.G. Milne pers. comm.). However, the amorphous and unstructured nature of the antigenic material in these tests precludes assignment of a definite MLO origin to such material and heavy reliance is necessarily placed on the specificity of the antibody probe for MLO antigens only.

Ultraviolet fluorescence microscopy

A major advantage of fluorescent light microscopy over electron microscopy is the greater speed with which samples can be examined, and also the larger amount of tissue in the field of view. The latter point is particularly important because of the frequently erratic distribution of MLOs in the vascular tissue.

Although the DAPI procedure does not involve the use of antibody diagnostic reagents the speed and sensitivity of the procedure justify its inclusion in this chapter on rapid detection methods (see Appendix 4). It may also be used in combination with fluorescent antibody stains. The primary application of this method is in detecting the presence of MLOs in sections of fresh or fixed vascular tissue from infected plants.

Immuno-fluorescence microscopy

The DAPI method is extremely simple and rapid and is the preferred method for rapid detection of MLO in phloem tissue. However, it is non-specific and cannot differentiate among MLOs. Such identification may be achieved by incubating tissue sections with specific antibody–fluorescein isothiocyanate (FITC) conjugates prior to examination by fluorescence microscopy (Appendix 5). This method may be combined with the DAPI technique both to detect and to identify MLOs present in tissue sections. The excitation and barrier filters used for DAPI and FITC are different so that by simply changing filters, the fluorescence contributed by each one may be assessed separately in double-stained sections (Fig. 2.3, facing page 50).

Enzyme-linked immunosorbent assay (ELISA)

Details of various ELISA procedures and protocols are presented in Chapter 1 so it is proposed to deal with only those aspects specifically relating to the detection and identification of MLOs here. Although direct assays can be employed, in which the detecting antibody is coupled directly with the enzyme label, most workers have found an indirect assay to be more reliable and sensitive, particularly when using polyclonal antisera to detect MLO antigens. As outlined above antigens may be adsorbed directly to the microplate or captured via a specific antibody previously adsorbed to the plate. The latter method is recommended wherever possible as it usually produces stronger ELISA values because the capture antibody traps a greater amount of antigen. The method used in my laboratory is the $F(ab')_2$ indirect ELISA described by Barbara & Clark (1982). $F(ab')_2$ fragments of rabbit polyclonal antibodies are coated onto the microplate in carbonate coating buffer and used to capture homologous MLO antigens from crude plant extracts or partially purified preparations. If MAbs are to be used as the detecting antibody, an appropriately diluted solution of MAb in PBS–Tween can be incubated with the immobilized antigens followed after rinsing by the goat anti-mouse (or anti-rat) immunoglobulin—enzyme conjugate. If polyclonal antibodies are to be used it may be advantageous to dilute the antibody in PBS—Tween containing a dilute extract of healthy plant tissue (e.g. 1 g leaf per 40 ml buffer) as an additional 'intra-well' cross-absorption treatment, to reduce background reactions. Such action is effective despite cross-absorbing the antiserum (Appendix 2) as it is not always possible to fully remove all anti-plant antibodies by the relatively crude cross-absorption treatment.

Final comments

Development of methods for detecting and identifying MLOs lag several years behind those available for other plant pathogens. Such methods as are in use, although encompassing the same sophisticated techniques frequently employed in other areas of plant pathogen diagnosis, have not yet been applied to more than a handful of MLOs. Consequently little definitive information is available and MLO identification and nomenclature is still reliant on conventional biological tests. However, the picture is beginning to change and several laboratories are engaged in eliciting information about the characteristics of MLOs. The methods described here should be regarded only as guides to some of the diagnostic possibilities that have been explored and found to be useful in my laboratory. Many other procedures are being investigated or used in other laboratories. It is important to recognize that because of the dearth of definitive information on MLOs any technique that adds significantly to the slowly growing fund of knowledge is bound to be of use in the detection and identification of this group of intractable plant pathogens.

References

Barbara D.J. & Clark M.F. (1982) A simple indirect ELISA using F(ab')₂ fragments of immunoglobulin. *Journal of General Virology* **58**, 315–22.

Caudwell A., Meignoz R., Kuszala C., Larrue J., Fleury A. & Boudon E. (1982) Purification serologiques et observation ultramicroscopique de l'agent pathogene (MLO) de la flavescence dorée de la vigne dans les extraits liquides de fèves (*Vicia faba* L.) malades. *Comptes Rendus des Séances de la Société de Biologie* **176**, 723–9.

Clark M.F., Davies D.L. & Barbara D.J. (1983) Production and characteristics of antisera to *Spiroplasma citri* and clover phyllody-associated antigens derived from plants. *Annals of Applied Biology* **103**, 251–9.

Clark M.F., Morton A. & Buss S.L. (1989) Preparation of mycoplasma immunogens from plants and a comparison of polyclonal and monoclonal antibodies made against primula yellows MLO-associated antigens. *Annals of Applied Biology* **114**, 111–24.

Clark M.F., Davies D.L., Barbara D.J. & Markham P.G. (1986) Serological comparisons among mycoplasma-associated antigens from experimentally infected plants. *Acta Horticulturae* **193**, 333–4.

Clark M.F., Davies D.L., Buss S.L. & Morton A. (1988) Serological discrimination among mycoplasma-like organisms using polyclonal and monoclonal antibodies. *Acta Horticulturae* **235**, 107–13.

Davis R.E., Lee I.-M., Dally E.L., Dewitt N. & Douglas S.M. (1988) Cloned nucleic acid hybridisation probes in detection and classification of mycoplasmalike organisms (MLOs). *Acta Horticulturae* **234**, 115–22.

Doi Y., Teranaka M., Yora K. & Asuyama H. (1967) Mycoplasma or PLT group-like microorganisms found in the phloem elements of plants infected with mulberry dwarf, potato witches'-broom, aster yellows, or Paulownia witches'-broom. *Annals of the Phytopathological Society of Japan* **33**, 259–66.

Hobbs H.A., Reddy D.V.R. & Reddy A.S. (1987) Detection of a mycoplasma-like organism in peanut plants with witches' broom using indirect enzyme-linked immunosorbent assay (ELISA). *Plant Pathology* **36**, 164–7.

Ishiie T., Doi Y., Yora K. & Asuyama H. (1967) Suppressive effects of antibiotics of tetracycline group on symptom development of mulberry dwarf disease. *Annals of the Phytopathological Society of Japan* **33**, 267–75.

Jiang Y.P. & Chen T.A. (1987) Purification of mycoplasma-like organisms from lettuce with aster yellows disease. *Phytopathology* **77**, 949–53.

Kirkpatrick B.C., Stenger D.C., Morris T.J. & Purcell A.H. (1987) Cloning and detection of DNA from a non-culturable plant pathogenic mycoplasma-like organism. *Science* **238**, 197–200.

Lin C.P. & Chen T.A. (1986) Comparison of monoclonal antibodies and polyclonal antibodies in detection of the aster yellows mycoplasmalike organism. *Phytopathology* **76**, 45–50.

McCoy R.E., de Leeuw G.T.N., Marwitz R. *et al.* (1987) Plant diseases associated with mycoplasma-like organisms. In Whitcomb R.F. & Tully J.G. (eds) *The Mycoplasmas* Vol. 5, Chapter 11. Academic Press, New York.

Seemüller E. (1976) Investigations to demonstrate mycoplasma-like organisms in diseased plants by fluorescence microscopy. *Acta Horticulturae* **67**, 109–12.

Sinha R.C. (1979) Purification and serology of mycoplasma-like organisms from aster yellows-infected plants. *Canadian Journal of Plant Pathology* **1**, 65–70.

Sinha R.C. & Benhamou N. (1983) Detection of mycoplasma-like organism antigens from aster yellow-diseased plants by two serological procedures. *Phytopathology* **73**, 1199–202.

Sinha R.C. & Chiykowski L.N. (1984) Purification and serological detection of mycoplasmalike organisms from plants affected by peach eastern X-disease. *Canadian Journal of Plant Pathology* **6**, 200–5.

Appendix 1: Preparation of MLO immunogen from infected plant tissue

Materials
- 0.1 M Phosphate buffer (PB), pH 7
- Phosphate buffered saline (PBS)
- Glycine magnesium (GM) buffer:
  0.3 M glycine–NaOH, pH 8
  0.02 M $MgCl_2$

Sucrose	50 g l^{-1}
Sodium mercaptoacetate	2 g l^{-1}
- PEG, mol. wt 6000
- BSA

Method   1   Wash about 15–20 g fresh tissue for 60 min under running tap water. All subsequent steps should be carried out at 0–4°C.
2   Blot dry, remove and discard non-vascular leaf lamina tissue.
3   Triturate tissue with four volumes of ice-cold GM buffer.
4   Squeeze the homogenate through cheesecloth, clarify at 2000 g for 5 min, then at 8000 g for 20 min.
5   Add PEG at 60 g l^{-1} and leave the extract on ice for 1 h for any precipitate to form. Centrifuge at 12 000 g for 40 min.
6   * Resuspend pellets with one tenth the original volume of GM buffer.
7   To aid resuspension and to remove residual PEG the preparation should be passed through a column of Sepharose 4B (Pharmacia Ltd) (bed volume at least 4× volume of resuspended preparation) equilibrated in GM buffer.
8   Clarify at 2000 g for 5 min.
9   Incubate the preparation with 5% (v/v) rabbit anti-plant antiserum for approximately 1 h to reduce contamination by plant antigens.
10   Clarify at 2000 g for 5 min.
11   Filter through a small column of protein A-agarose to remove residual antigen-bound and unreacted rabbit immunoglobulins.
12   Centrifuge for 1 h at 63 000 g.
13   Resuspend the pale green pellet in PBS, mix with adjuvant and inject.

Appendix 2: Cross-absorption of antiserum

Method   • Wash 10 g vascular tissue from healthy plant leaves for 60 min under running tap water.
• Blot dry.
• Triturate tissue with four volumes of ice-cold GM buffer (see Appendix 1).
• Squeeze the homogenate through cheesecloth, clarify at 2000 g for 5 min, then at 8000 g for 20 min.
• Add PEG to 60 g l^{-1} and leave the extract on ice for 1 h for any precipitate to form. Centrifuge at 12 000 g for 40 min.
• Resuspend pellet in one tenth the original volume of PBS. Add 0.2 g l^{-1} sodium azide and 0.1 g l^{-1} freshly prepared phenyl methyl sulfonyl fluoride (or other enzyme inhibitor).
• Add 0.5 ml preparation to antiserum. Incubate for 30–60 min at 37°C.
• Centrifuge at 2000 g for 10 min. Discard the pellet.

Repeat the last two steps until no further flocculation occurs, then incubate the antiserum at 4°C for 24 h and finally centrifuge at 8000 g for 10 min. Antibodies should be extracted and purified from the cross-absorbed preparation as soon as possible to minimize enzyme-mediated degradation.

---

* At this stage the pellets may be resuspended in PBS and injected directly, or further treated to remove additional plant components. Injection of the re-suspended PEG pellet has been found advantageous in preparing antisera where additional steps have resulted in unacceptable loss of antigenic activity.

Appendix 3: Immunogold labelling of MLO in glutaraldehyde-fixed tissue sections (R.G. Milne & R. Lenzi, pers. comm.)

Method
- Harvest tissue and fix in 2.5% glutaraldehyde in 0.1 M PB, pH 7.0 for 1 h at room temperature.
- Excise vascular tissue and slice into pieces about 0.2–0.5 mm in length.
- Rinse pieces in PB and incubate in MAb diluted appropriately in PB containing 1% BSA and 0.1% Tween 20, for 2 h at room temperature.
- Rinse in PB–Tween and then incubate in gold-labelled goat-anti-mouse antibody (5 nm) diluted according to the manufacturer's recommendations in PB–BSA–Tween for 2 h at room temperature.
- Rinse in PB–Tween and fix in 0.1% osmium tetroxide in PB for 1 h.
- Proceed with routine embedding of specimens in Epon for sectioning and electron microscope examination.

Appendix 4: Detection of MLOs in plant tissue using DAPI (Seemüller, 1976)

Method
- Collect samples or pieces of vascular tissue from plants to be examined. These may be examined immediately, or preserved for later examination by storage at 4°C in 0.1 M PB, pH 7.0, containing either 5% glutaraldehyde or 4% formaldehyde.
- Prepare 15–30 µm thick sections using a freezing microtome and suitable cryoembedding medium. Transfer sections to a clean glass slide with a fine brush.
- Flood the sections with 0.1 M PB containing 0.1 mg l^{-1} DAPI and leave for 5–10 min at room temperature.
- Place a coverslip over the sections and blot firmly with filter paper to remove excess stain. Coverslips may be ringed with nail varnish for temporary storage of specimen.
- Examine sections with a high efficiency epifluorescent microscope, using an excitation filter wavelength in the range 300–400 nm and a barrier filter transmitting light above 400 nm. An epifluorescence oil immersion objective is necessary for examining sections at high magnification.

In longitudinal sections at a magnification of ×600, MLOs appear as small, bright blue dots in the phloem cells of infected tissue which, with practice, can be readily distinguished from fluorescence of other DNA-containing organelles such as mitochondria or nuclei.

Appendix 5: Immunofluorescent staining of MLOs within plant tissue

Method
- Purify antibodies to be conjugated and prepare an antibody solution at 2 mg ml^{-1} in 0.1 M sodium carbonate buffer, pH 9.5 (dialyse to remove inhibitory salts).
- Freshly prepare a solution of the fluorochrome, e.g. FITC at 1 mg ml^{-1} in dimethyl sulphoxide (DMSO).
- Add FITC solution dropwise to antibody, 50 µl FITC ml^{-1} for polyclonal antibodies or 20 µl for MAbs.

● Mix and allow to stand in the dark for 2 h at room temperature, or for 8 h at 4°C.

● Separate unbound dye from the conjugate by gel filtration on Sephadex G-25 in PBS. Use a bed volume of 20× the sample volume to ensure adequate separation. The conjugate should be easily visible as the first fluorescing material to elute from the column.

● Store the conjugate in a lightproof container at 4°C. A small amount of BSA (e.g. $5-10 \mu g \, ml^{-1}$) may be added to help stabilize the conjugate.

● Incubate tissue sections for 30 min in a moist chamber with a range of conjugate dilutions to establish appropriate conditions for use. Excitation and barrier filters will vary according to the fluorochrome employed.

If over- or under-staining is observed the conjugation procedure should be repeated with appropriate adjustments to the fluorochrome: antibody ratio.

3 Detection of plant-invading fungi by monoclonal antibodies

F.M. DEWEY

Introduction

There are some important fungal diseases, particularly those that are soil-borne and which affect the stem base, where diagnosis is difficult and early treatment is essential. Development of quick diagnostic assays for such diseases would help reduce both crop losses and unneccessary spraying. Immunological techniques, which are rapid, sensitive and specific, could be used for the rapid detection of fungal pathogens and diagnoses of the diseases which they cause. They would also be helpful for research purposes, epidemiological studies, quantitative measurements and identification of non-fruiting mycelia which are common on plant roots.

Unfortunately the development of immunodiagnostic techniques and the use of monoclonal antibodies (MAbs), in particular, to detect specific fungi in diseased plants has been slow. Progress in this area contrasts markedly with the relatively rapid and successful development of serological assays incorporating both polyclonal and MAbs for plant viruses. There are several reasons for this slow progress. There was less urgency to develop non-visual diagnostic methods because several fungal pathogens, such as rusts and mildews, produce characteristic spores that can be seen with the light microscope and used for identification. Also many fungal pathogens can be induced to grow out from surface sterilized diseased material into culture and sporulate. Furthermore, selective media have been developed for some fungi that will discourage the growth of quick-growing non-invasive fungi, thereby facilitating the isolation of specific pathogens. However, these latter techniques clearly are not suitable for obligate pathogens, fungi that do not produce spores *in vitro* or those that are slow growing. They are also inappropriate where time and expertise are limited.

Probably the most significant reason for the lack of progress in development of fungal immunodiagnostics has been the difficulty of raising antisera that are species-specific. Antisera to fungal spores, hyphal fragments, solubilized extracts or concentrates of culture filtrates, when tested by immunodiffusion, may appear to be specific but when tested by more sensitive methods such as immunofluorescence (IF) or enzyme-linked immunosorbent assay (ELISA) invariably prove to be non-specific (Chard *et al.*, 1985 a,b; F.M. Dewey unpublished results). Such antisera cross-react widely with related species, species from unrelated genera and host molecules (Mohan & Ride, 1982; Gendloff *et al.*, 1983; Kough *et al.*, 1983; Musgrave, 1984; Aldwell *et al.*, 1985; Chard *et al.*, 1985 a,b; Hardham *et al.*, 1986; Polonelli *et al.*, 1986; Kaufman & Standard, 1987; Notermans *et al.*, 1987; Dewey & Brasier, 1988; Dewey *et al.*, 1989 a,b; Mohan, 1989). Nevertheless, Barker & Pitt (1988) produced a polyclonal antiserum which, despite some cross-reactions with unrelated fungi, was specific enough to detect the pathogen (*Colletotrichum* sp.) which causes leaf curl of anemone in infected corms. Several workers have attempted to increase antiserum specificity by cross-absorption with un-related fungal species. These methods generally have improved specificity but do not eliminate all cross-reactions and most workers have found that cross-absorption decreases the titre considerably (Chard *et al.*, 1985 b; Gerik *et al.*, 1987). Other workers have attempted to dilute out non-specific antibodies (Musgrave, 1984) or to block non-specific interactions with bovine serum albumin (BSA) (Willingdale & Mantle, 1987). Cross-reactions appear to be less problematical when competition assays are used (Kitagawa *et al.*, 1989).

The site and nature of species, subspecies and race- or isolate-specific antigens is still not known. However, several workers have demonstrated, by immunodiffusion, immunoelectrophoresis and electroblot immunoassay, that antisera raised against hyphal fragments or soluble extracts of a fungus contain antibodies with high specificity towards some components of that original fungus which will not cross-react with extracts of other fungi (Dewey *et al.*, 1984; Chard *et al.*, 1985 a; Mohan, 1989). However, attempts at concentrating such antibodies from poly-clonal antisera have generally failed.

Monoclonal antibodies

It is possible to obtain MAbs by hybridoma technology (Kohler & Milstein, 1975), which uses cell culture techniques to fuse lymphocytes with continuously growing myeloma cells producing hybridomas which secrete antibodies (MAbs) of defined specificity. It is therefore surprising that there have been, apparently, relatively few attempts to raise MAbs to fungi including medically important species (Polonelli *et al.*, 1986; Kaufman & Standard, 1987). There are only a few published reports of MAbs raised against plant-related fungi (Table 3.1). The first

Table 3.1 Monoclonal antibodies raised against plant-invading fungi

Fungus	Nature	Immunogen (amount per injection)	Freund's adjuvant	Route Early	Route Booster	Screening method	Specificity	MAb class	Reference
Fusarium oxysporum f. sp. *lycopersici*	Spores	10^7	+	i.p.	i.p.	ELISA-spores	Genus	NG	Ianelli *et al.* (1983)
Tilletia sp.	Spores Spore extracts	3 mg –	– –	i.p. i.p.	i.v. i.v.	ELISA-spores ELISA-spores	NSG	IgM,	Banowetz *et al.* (1984)
Phytophthora cinnamomi	Zoospores	2×10^7	–	i.p.	i.p.	IF-zoospores	NSG	NG	Hardhan *et al.* (1986)
Sirococcus strobilinus	Hyphal fragments	1 mg	+	i.p.	i.v.	ELISA-sol. Ag	NS Genus	IgM, IgG	Mitchell & Sutherland (1986)
Glomus occultum	Spores	50 000	–	i.p.	i.p.	ELISA-spores	Species	IgG	Wright *et al.* (1987)
Phytophthora megasperma f. sp. *glycinea*	Culture filtrate	0.15 mg protein	+	i.p. intrasplenic	NG	ELISA-sol. Ag Western blot	NSG	IgM, IgG, IgA	Wycoff *et al.* (1987)
Fusarium oxysporum f. sp. *cubense*	Conidia Hyphal fragments	8×10^7 40 µg	–			IF	NSG Race specific	IgM	Wong *et al.* (1988)
Armillaria mellea	Hyphal fragments	NG	+/–	i.p.	NG	ELISA-hyphal fragments	NSG	ND	Fox & Hahne (1989)
Ophiostoma ulmi	Hyphal fragments	3 mg	+	i.p.	i.v.	ELISA-sol. Ag	Species & genus	IgG, IgM	Dewey *et al.* (1989 a)
Humicola lanuginosa	SW	150 µg protein	–	i.p.	i.p.	ELISA-SW	Genus*	IgM	Dewey *et al.* (1989 b)
Penicillium islandicum	SW	500 µg protein	–	i.p.	i.p.	ELISA-SW	Species	IgG	Dewey *et al.* (1990)
Pythium	Zoospores	NG	NG	NG	NG	IF-zoospores	Zoospores & cysts, genus & species	NG	Callow *et al.* (1987) Estrada-Garcia *et al.* (1989)

* Tested against species from 18 unrelated genera; cross-reacts with 2.

ELISA = enzyme-linked immunosorbent assay; i.p. = intraperitoneal; i.v. = intravenous; NG = not given; ND = not determined; S.W. = surface washings; NSG = not tested against other species or genera; NS = not tested against other species; sol. Ag = soluble antigens.

studies were done by Ianelli *et al.* (1983) who attempted to raise MAbs that would differentiate form species (f. sp.) of *Fusarium oxysporum*. They, like Banowetz *et al.* (1984), only obtained MAbs that would give quantitative differences. Banowetz *et al.* (1984) tried to raise MAbs that would specifically differentiate teliospores of two species of wheat bunt fungi.

Most of the MAbs raised so far have not been raised for diagnostic purposes. However, some, such as the species-specific MAbs raised against *Phytophthora cinnamomi* (Hardham *et al.*, 1986), *Phytophthora megasperma* var. *glycinea* (Wycoff *et al.*, 1987) and *Pythium aphanidermatum* (Callow *et al.*, 1987; Estrada-Garcia *et al.*, 1989), clearly have diagnostic potential. It is unfortunate that the species-specific MAbs for *Pythium* and *Phytophthora* raised by Hardham *et al.* (1986) and Estrada-Garcia *et al.* (1989) only differentiate species on the basis of zoospores. Specific assays for the detection of mycelia in infected plants or soil would be more useful. Wong *et al.* (1988) have raised a MAb to the banana wilt fungus, *Fusarium oxysporum* f. sp. *cubense* that will differentiate the thick-walled chlamydospores of strain 4 from those of strains 1 and 2 by IF. We have raised MAbs that are specific to the Dutch elm disease pathogen *Ophiostoma (Ceratocystis) ulmi* (Dewey *et al.*, 1989 a), the eyespot pathogen of cereals *Pseudocercosporella* (Dewey, 1988) and two fungi involved in post-harvest deterioration of rice grains *Humicola lanuginosa* and *Penicillium islandicum* (Dewey *et al.*, 1989 b, 1990).

The following guidelines for the development of MAbs to fungi are based on our own experiences and those of recent workers, particularly Mitchell (1988), who have successfully developed immunodiagnostic assays for plant-invading fungi.

Assays for the routine detection of fungal pathogens

Before starting to raise MAbs to specific fungi, it is important to identify the goals of the project. The versatility of the MAbs selected and the types of assays in which they will function can be limited by the choice of immunogen and method of screening hybridoma cell culture supernatant fluids (CSF). Some MAbs will recognize their antigens equally well when used in ELISA, IF, dot-blots or to probe Western blots of polyacrylamide gels, but others will only function in one type of assay. It is therefore important to determine at the outset the requirements of the final detection assay and hence the goals of the project. The following points should be taken into consideration.

1 Facilities available to do the assays — field or laboratory.
2 Number of samples to be tested per day.
3 The need to produce results quickly.

Labels in figure (a):
External phloem
Xylem vessels
Internal phloem

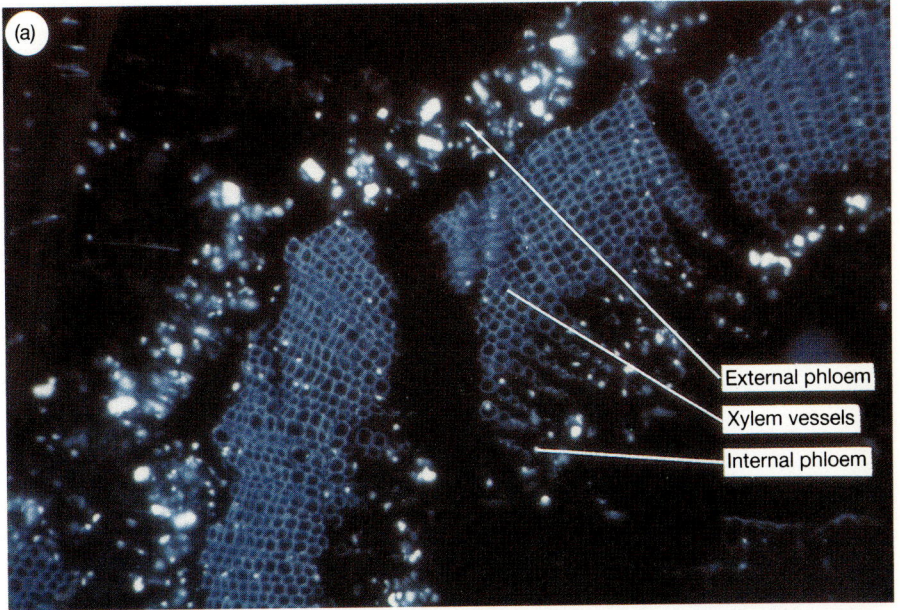

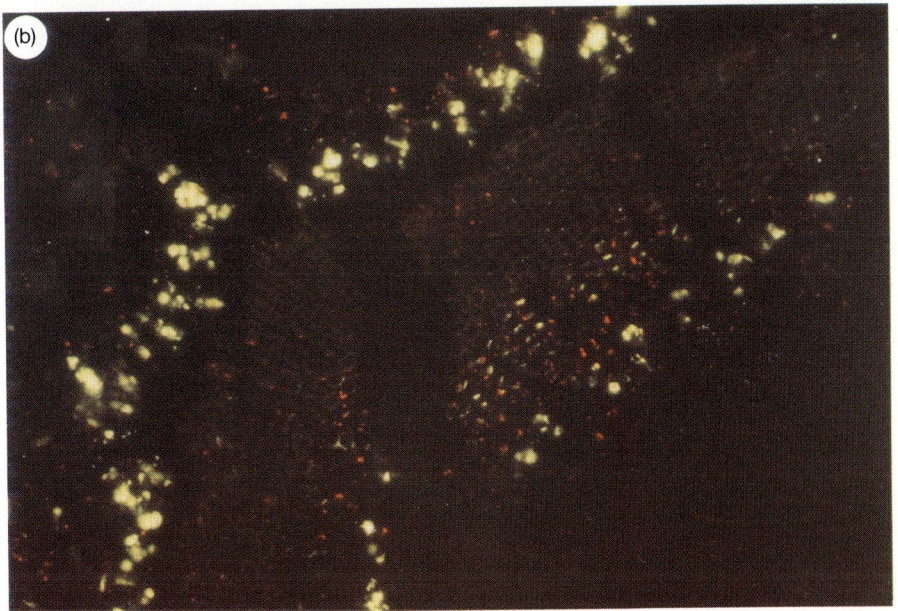

Fig. 2.3 Ultraviolet fluorescence microscopy of a transverse section of part of the stem of *C. rosea* infected with primula yellows MLO. The section was double-stained with (a) DAPI and with (b) FITC-labelled EM1 monoclonal antibody to primula yellows MLO. Filter combinations were (a) excitation = 340–380 nm barrier = 430 nm (for DAPI stain); (b) excitation = 450–490 nm barrier = 515 nm (for FITC-labelled antibodies).

4 Ease of extraction of pathogen from host.

If MAbs are being raised for laboratory-based detection purposes, with specificity at the species level, then an ELISA system using soluble antigens is recommended (Mitchell & Sutherland, 1986; Dewey *et al.*, 1989 a). If, however, a quick 'user-friendly' field assay is needed, a dip-stick or dot-blot method of detection using soluble, easy-to-extract antigens would be more appropriate (Dewey *et al.*, 1989 b; Mitchell, 1988). We have shown that species-specific MAbs raised against saline surface washings of fungi can be used equally well in ELISA or dip-stick assay systems (Dewey *et al.*, 1989 b, 1990). The use of insoluble material, such as hyphal fragments or spores, is not recommended for ELISA. Banowetz *et al.* (1984) using antigens fixed to microtitre plates with poly-L-lysine or glutaraldehyde were unable to obtain MAbs that differentiated between two species of *Tilletia*. Likewise, J.M. Booth and J.G. White (pers. comm.) were unable to obtain species-specific MAbs using hyphal fragments of *Pythium* fixed to microtitre wells. However, Hardham *et al.* (1986) and Estrada-Garcia *et al.* (1989) have had considerably more success in raising MAbs that are specific to zoospores of *Phytophthora* and *Pythium* spp. using zoospores fixed to microtitre wells with poly-L-lysine.

IF assays are time consuming and not recommended where large numbers of samples have to be screened. However, these types of assays may be the only way to detect fungal propagules in soil. Furthermore, Wong *et al.* (1988), who raised a MAb that differentiated thick-walled chlamydospores of race 4 of *Fusarium oxysporum* f. sp. *cubense* from races 1 and 2, argue that this MAb could not have been selected by an ELISA type assay.

Selection and preparation of immunogens

The choice of immunogen for the production of anti-fungal MAbs is critical and not easy to make. The potential range of MAbs produced in a fusion reflects only the nature and range of antigens presented to the animal. In general, the higher the percentage of species-specific molecules in the immunogen the higher the percentage of clones secreting MAbs that are specific. Until we know if and where species-specific antigens are located we can only either inject a mixture of antigens and select the few hybridomas that recognize the specific molecules (Mitchell & Sutherland, 1986) or first separate fractions containing specific molecules by high performance liquid chromatography (HPLC) or electrophoresis (MacDonald *et al.*, 1989). The former approach, possibly improved by first concentrating fungal proteins, involves a lot of time screening hundreds of CSF, but the screening itself is relatively easy and the results are clear cut. The

latter method involves considerable effort in the fungal purification steps before immunization and has several other disadvantages. Firstly, it is probable that the molecules of choice,
particularly if they are glycoproteins or carbohydrates, will
share many common epitopes and thus purification of a particular molecule does not ensure the specificity of the MAb
raised against it. Secondly, by pre-selecting one particular
antigen purified *in vitro* we are not necessarily selecting an
antigen that is produced in significant quantity *in vivo* (Dewey
et al., 1989 a).

We have used both mycelial homogenates and phosphate
buffered saline (PBS) surface washings of cultures to raise
species-specific MAbs with relatively little difficulty, with the
unexplained exception of the cereal eyespot pathogen *Pseudocercosporella herpotrichoides* (Dewey, 1988). Surface washings
were prepared by brief wash with 1 ml PBS of a mycelial colony
grown on a solid slant culture medium. The surface of the
colony was stroked with a sterile plastic pipette tip and the
washings removed by suction. After centrifugation to remove
cell debris the wash was injected directly (without Freund's
adjuvant) into the peritoneum of a mouse (300 µl per injection)
(Dewey *et al.*, 1989 b; Appendix 1).

A summary of the range of immunogens, amounts and routes
of immunization that have been successfully used to raise
species-specific MAbs to fungi is given in Table 3.1. Further
general information on the range of immunogens that have
been used to raise MAbs, the amounts injected and injection
schedules are given by Harlow & Lane (1988). Injections into
the peritoneum are the most common, easiest to perform and
can be used for either particulate or non-particulate material
with or without Freund's adjuvant. If booster injections are
also given by the intraperitoneal route, it is worth noting that
Harlow & Lane (1988) recommend leaving 5–7 days before the
fusion. Immunogens containing particulate matter should not
be injected into the tail vein. It is not yet clear which route for
the booster injection (tail vein or peritoneum) is most effective
in raising antigens to specific fungal molecules. Neither is it
clear whether the use of Freund's adjuvant is helpful.

Fusion protocol

We have raised hybridoma cell lines by centrifuging splenocytes
and myeloma cells together in the presence of 30% polyethylene
glycol (PEG) mol. wt 1000, using the method described by
Harlow & Lane (1988) which, in turn, is based on the methods
of Gefter *et al.* (1977) and Kennett (1978). Splenocytes (about
10^8 cells) from an immunized mouse were released by teasing

the spleen apart with sterile needles in the presence of 10 ml of serum-free medium (RPMI). Both splenocytes and myeloma cells (about 10^7, cell line SP2/0-Ag-14) were washed by centrifuging at 250 g for 5 min in serum-free medium, then pooled and centrifuged together at 250 g for 5 min. Fusion was brought about by gently resuspending the combined cells in 0.2 ml of 30% (v/v) PEG in serum-free medium for 5 min and then centrifuging for 2 min at 500 g. The PEG solution was then carefully sucked off and the cells gently resuspended in 5 ml serum-free medium followed by further dilution with 45 ml medium containing 20% foetal bovine serum. The cell suspension was distributed over ten 96-well microtitre culture plates at 50 µl per well. The following day 50 µl of medium containing double-strength selective agents hypoxanthine and azaserine was added to each well.

Screening procedure for hybridoma cell culture supernatant fluids

The screening procedure can be developed using antisera from test bleeds after the second or third injections and should be used to determine the titre of the antisera from mice before the fusion. Titres of antisera to fungi from mice having had three or more injections are normally very high, ranging from 1 in 60 000 to 1 in 250 000 when tested by ELISA, but much lower, approximately 1 in 40, when tested by IF. As indicated earlier the specificity of fungal antisera is generally low but it does vary from animal to animal and should be determined in order to select the most favourable mouse. Antibody specificity can be determined by testing CSF or antisera against polystyrene microtitre wells coated with antigens from a range of fungi (for details see Appendices 1 and 2). A two- or three-step screening schedule is recommended. First, all CSF must be screened against the immunogen. Second, those that are positive are then screened against antigens from fungi with which the pathogen is most commonly confused and against extracts from host material. A high degree of specificity in tests on laboratory-grown fungal cultures does not ensure that the MAb will not react with extracts of host plant material (Dewey *et al.*, 1989 a). Finally, extensive testing should be done against all related and unrelated species which may be present in the sample material. Details of ELISA, IF and dot-blot screening procedures for fungi are given in Appendices 2, 3 and 4.

Development of detection assay

The most sensitive and specific assays are probably those using the classical sandwich ELISA techniques that employ either two MAbs, one as the capture antibody and the other conjugated to an enzyme as the detector antibody, or polyclonal antibody from another animal species as the capture antibody. However, with the exception of a few commercially available kits such as those produced by Agri-Diagnostics Associates (Miller *et al.*, 1988, this volume) almost no MAb sandwich ELISA kits have been developed to detect plant-invading fungi. The majority of fungal detection assays remain relatively unsophisticated depending on direct antigen coating of wells or membranes by fungal extracts or solutes which have diffused from the sample (Appendix 1) (Mitchell & Sutherland, 1986; Dewey *et al.*, 1989 a,b). Such methods can have distinct advantages as, for example, in the detection of fungi on the outer surfaces of rice grains. By this method a number of grains can be tested individually with relative ease. Grains are soaked individually overnight in PBS in microtitre plate wells or Eppendorf tubes, which allows the soluble material, containing fungal antigens, to coat the wells or dip-stick directly. Frequently, it is the knowledge of the numbers of infected grains or seeds that is more important than an average figure for fungal contamination (Dewey *et al.*, 1989 b, 1990; Mitchell, 1988).

ELISA detection assays done in a laboratory are undoubtedly the most informative but are unsuitable for field-based assays. It is probable therefore that membrane-bound detection systems will gain in popularity. Agri-Diagnostics Associates have developed an excellent antibody-coated nitrocellulose membrane detection system that can be used to detect *Rhizoctonia*, *Pythium*, *Sclerotinia* and *Phytophthora* in turfgrass and/or other crops (Miller *et al.*, 1988, this volume). The kits are specifically designed for 'speed (10 min), simplicity and sensitivity'. They include negative and positive controls and have a sensitivity threshold of $1\,\mu g\,ml^{-1}$ and $0.25\,\mu g\,ml^{-1}$ protein for *Rhizoctonia* and *Phytophthora* respectively (Miller *et al.*, this volume). Mitchell (1988) has developed a similar antigen-capture dot-immunoassay system to detect *Sirococcus strobilinus* in extracts from infected spruce seeds. Surface-sterilised seeds (500) were homogenized with bicarbonate buffer (5 ml) containing proteolytic enzyme inhibitors which were then centrifuged and spotted onto the membrane in $1\,\mu l$ droplets. Using a mixture of four MAbs they found the dot-immunoassay system to be more sensitive than antigen-capture ELISA tests. Threshold detection levels of $1-5\,ng$ fungal protein were obtained.

We have developed a simple dip-stick and dot-blot assay detection system for *H. lanuginosa* and *P. islandicum* in rice grains using a membrane made of polyvinylidene difluoride (PVDF) (Immobilon P from Millipore) (Dewey *et al.*, 1989 b,

1990). Fungal antigens bind strongly to the PVDF membrane surface. An immunogold conjugate was used as the detector antibody system and the reaction was enhanced with silver (Appendix 3). The sensitivity of this system has not been directly compared to ELISA tests but it seems to be somewhat less sensitive. However, the big advantage of this system is that samples can be placed on the membrane and stored for a long period before testing.

Problems with non-specific reactions are common in any assay system involving extracts of plant material. Most workers routinely block antigen-coated wells or membranes with BSA or solutions of non-fat milk powders (Appendix 3). It is also important to check that the anti-mouse antibody or other detector−antibody enzyme conjugate does not react with host antigens or fungal molecules. Some antibody−enzyme conjugates will react with fungi non-specifically giving low levels of false positives. So it is important to check that endogenous enzyme levels in the host extract are not contributing to the final results. Such problems have been overcome by dilution, heating extracts, use of inhibitors at an early stage or by using immunogold conjugates (Dewey, 1988).

Latex agglutination assays have been developed to detect fungal contaminants in food but to our knowledge such methods have not been developed for fungal plant pathogens (Notermans et al., 1987). However, the speed and ease of use of latex agglutination tests is attractive and warrants some investigation.

Conclusion

The need for commercial detection assays for plant diseases and pathogens is clear, as is the potential of MAbs to form the basis of such tests, but at the present time only a few commercial kits exist and many aspects remain unexplored.

Several workers have shown that it is possible to raise species-specific MAbs to plant-invading fungi. Studies with Pythium and Phytophthora spp. indicate that it is easier to raise species-specific MAbs to zoospore membranes than to cell walls (Hardham et al., 1986; Callow, 1987; Estradia-Garcia et al., 1989). Experiences in raising MAbs to mycelial fragments or surface washings have been mixed; it appears to be easier with some fungi such as Ophiostoma ulmi (Dewey et al., 1989 a) and Sirococcus strobilinus (Mitchell & Sutherland, 1986) than with others, for example Fusarium spp. (Ianelli, 1983) and P. herpotrichoides (Dewey, 1988). It is probable that the latter fungi contain non-specific dominant antigens. Methods for blocking or avoiding these antigens have not yet been developed.

References

Aldwell F.E.B., Hall I.R. & Smith J.M.B. (1985) Enzyme-linked immunosorbent assays as an aid to taxonomy of the Endogonaceae. *Transactions of the British Mycological Society* **84**, 399−402.

Banowetz G.M., Trione E.J. & Krygier B.B. (1984) Immunological comparisons of teliospores of two wheat bunt fungi, *Tilletia* species using monoclonal antibodies and antisera. *Mycologia* **76**, 51−62.

Barker I. & Pitt D. (1988) Detection of the leaf curl pathogen of anemones in corms by enzyme-immunosorbent assay (ELISA). *Plant Pathology* **37**, 417−22.

Callow J.A., Estrada-Garcia M.T. & Green J.T. (1987) Recognition of non-self: the causation and avoidance of disease. *Annals of Botany* **60**, Suppl. 4, 3−14.

Chard J.M., Gray T.R.G. & Frankland J.C. (1985 a) Purification of an antigen characteristic for *Mycena galopus. Transactions of the British Mycological Society* **84**, 235−41.

Chard J.M., Gray T.R.G. & Frankland J.C. (1985 b) Use of an anti-*Mycena galopus* serum as an immunofluorescent reagent. *Transactions of the British Mycological Society* **84**, 243−9.

Dewey F.M. (1988) Development of immunological diagnostic assays for fungal plant pathogens. *Proceedings of the Brighton Crop Protection Conference − Pests and Diseases, 1988* pp. 777−86. British Crop Protection Council, Thornton Heath, Surrey.

Dewey F.M. & Brasier C.M. (1988) Development of ELISA for *Ophiostoma ulmi* using antigen coated wells. *Plant Pathology* **37**, 28−35.

Dewey F.M., Munday C.J. & Brasier C.M. (1989 a) Monoclonal antibodies to specific components of the Dutch Elm Disease pathogen *Ophiostoma ulmi. Plant Pathology* **38**, 9−20.

Dewey F.M., MacDonald M.M. & Phillips S.I. (1989 b) Development of monoclonal antibody-ELISA, -dot-blot and -dip-stick immunoassays for *Humicola lanuginosa* in rice. *Journal of General Microbiology* **135**, 361−74.

Dewey F.M., Barrett D.K., Vose I.R. & Lamb C.J. (1984) Immunofluorescence microscopy for detection and identification of propagules of *Phaseolus schweinitzii* in infested soil. *Phytopathology* **74**, 291−6.

Dewey F.M., MacDonald M.M., Phillips S.I. & Priestley R.A. (1990) Development of monoclonal-antibody-ELISA and dip-stick immunoassays for *Penicillium islandicum* in rice grains. *Journal of General Microbiology* **136**, 753−60.

Estrada-Garcia M.T., Green J.R., Booth J.M., White G. & Callow J.A. (1989) Monoclonal antibodies to cell surface components of zoospores and cysts of the fungus *Pythium aphanidermatum* reveal species-specific antigens. *Experimental Mycology*, **13**, 348−56.

Fox R.T.V. & Hahne K. (1989) Prospects for the rapid diagnosis of *Armillaria* by monoclonal antibody ELISA. In Morrison D.J. (ed.) *Proceedings of the Seventh International Conference on Roots and Butt Rots, Victoria, Canada*, pp. 458−69. Forestry Canada, Victoria.

Gefter M.L., Margulies D.H. & Scharff M.D. (1977) A simple method for polyethylene glycol-promoted hybridization of mouse myeloma cells. *Cell Genetics* **3**, 231−6.

Gendloff E.H., Ramsdell D.C. & Burton C.L. (1983) Fluorescent antibody studies with *Eutypa armeniaceae. Phytopathology* **73**, 760−4.

Gerik J.S., Lommel S.A. & Huisman O.C. (1987) A specific serological staining procedure for *Verticillium dahliae* in cotton root tissue. *Phytopathology* **77**, 261−6.

Hardham A.R., Suzaki E. & Perkin J.L. (1986) Monoclonal antibodies to isolate-, species-, and genus-specific components on the surface of zoospores and cysts of the fungus *Phytophthora cinnamomi. Canadian Journal of Botany* **64**, 311−21.

Harlow E.D. & Lane D. (1988) *Antibodies: A Laboratory Manual.* Cold Spring

Harbor Laboratory, Cold Spring Harbor, New York.

Ianelli D., Caparelli R., Mariziano F., Scala F. & Noviello C. (1983) Production of hybridoma secreting monoclonal antibodies to the genus *Fusarium*. *Mycotaxon* **17**, 523–32.

Kaufman L. & Standard P.G. (1987) Specific and rapid identification of medically important fungi by exoantigen detection. *Annual Review of Microbiology* **41**, 209–25.

Kennett R.H. (1978) Cell fusion. *Methods in Enzymology* **58**, 345–59.

Kitagawa T., Sakamoto Y., Furumi K. & Ogwra H. (1989) Novel enzyme immunoassays for specific detection of various *Fusarium* species. *Phytopathology* **79**, 162–5.

Kohler G. & Milstein C. (1975) Continuous culture of fused cells secreting antibody of predefined specificity. *Nature* **256**, 595–7.

Kough J., Malajcjuk N. & Linderman R.G. (1983) Use of the indirect immunofluorescent technique to study the vesicular–arbuscular fungus *Glomus epigaeum* and other *Glomus* species. *New Phytologist* **94**, 57–62.

MacDonald M.M., Dunstan R.H. & Dewey F.M. (1989) Detection of low-Mr glycoproteins in surface washes of some fungal cultures by gel-filtration HPLC and monoclonal antibodies. *Journal of General Microbiology* **135**, 375–83.

Miller S.A., Rittenburg J.H., Petersen F.P. & Grothaus G.D. (1988) Application of rapid, field-useable immunoassays for the diagnosis and monitoring of fungal pathogens in plants. *Proceedings of the Brighton Crop Protection Conference – Pests and Diseases, 1988* pp. 795–803. British Crop Protection Council, Thornton Heath, Surrey.

Mitchell L.A. (1988) A sensitive dot immunoassay employing monoclonal antibodies for detection of *Sirococcus strobilinus* in spruce seed. *Plant Disease* **72**, 664–7.

Mitchell L.A. & Sutherland J.K. (1986) Detection of seed-borne *Sirococcus strobilinus* with monoclonal antibodies in an enzyme-linked immunosorbent assay. *Canadian Journal of Forest Research* **16**, 945–8.

Mohan S.B. (1989) Cross-reactivity of antiserum raised against *Phytophthora* species and its evaluation of a genus detecting antiserum. *Plant Pathology* **38**, 352–63.

Mohan S.B. & Ride J.P. (1982) An immunoelectrophoretic approach to the identification of progressive and fluctuating isolates of the hop wilt fungus *Verticillium albo-atrum*. *Journal of General Microbiology* **128**, 255–65.

Musgrave D.R. (1984) Detection of an endophytic fungus of *Lolium perenne* using enzyme-linked immunosorbent assay (ELISA). *New Zealand Journal of Agricultural Research* **27**, 283–8.

Notermans S., Wieten G., Engel H.W.B., Rambouts R.M., Hoogerhout P. & van Boom J.H. (1987) Purification and properties of extracellular polysaccharide (EPS) antigens produced by different mould species. *Journal of Applied Bacteriology* **62**, 157–66.

Polonelli L., Castagnola M. & Morace G. (1986) Identification and serotyping of *Microsporum canis* isolates by monoclonal antibodies. *Journal of Clinical Microbiology* **23**, 609–15.

Willingdale J. & Mantle P.G. (1987) Interaction between *Claviceps purpurea* and *Tilletia caries* in wheat. *Transactions of the British Mycological Society* **89**, 145–53.

Wong W.C., White M. & Wright I.G. (1988) Production of monoclonal antibodies to *Fusarium oxysporum* f. sp. *cubense* race 4. *Letters in Applied Microbiology* **6**, 39–42.

Wright S.F., Mortorn J.B. & Sworobuk J.E. (1987) Identification of a vesicular–arbuscular mycorrhiza fungus by using monoclonal antibodies in an enzyme-linked immunosorbent assay. *Applied and Environmental Microbiology* **53**, 2222–5.

Wycoff K.L., Jellison J. & Ayers A.R. (1987) Monoclonal antibodies to glycoprotein antigens of a fungal plant pathogen, *Phytophthora megasperma* f. sp. *glycinea. Plant Physiology* **85**, 508–15.

Appendix 1: Preparation of antigenic material

Soluble antigens from lyophilized mycelia
- Suspend lyophilized mycelial fragments in phosphate buffered saline (PBS — see Appendix 2) in an Eppendorf tube for 30 min at 4°C (20 mg ml^{-1} recommended).
- Centrifuge in a benchtop centrifuge (13 250 g) for 2 min to remove cell debris and dilute in PBS to appropriate concentration (1 mg ml^{-1} recommended).

Soluble antigens from PBS surface washings
- Wash surface of a slant culture with 1 ml PBS. Stroke surface of colony with a sterile plastic pipette tip to help 'wet' the surface (which is often hydrophobic).
- Remove washing solution by suction, place in an Eppendorf tube and centrifuge at 132 250 g for 2 min.
- Suck off supernatant and dilute (usually 1 in 20) with PBS before coating wells.

Insoluble antigens

Passive adsorption: Leave suspensions of fungal spores and small fragments of hyphae in PBS overnight in microtitre wells. Particles separate out and some adhere to the surface but the amount that remains fixed after the various washing procedures will vary. Passive adsorption is therefore inadvisable.

Poly-L-lysine: This has proved most popular for fixing mammalian cells to solid surfaces (Harlow & Lane, 1988).
- Incubate 1 mg ml^{-1} solution of poly-L-lysine (average mol. wt 400 000) in distilled water in the wells of a microtitre plate.
- Wash the wells with water prior to coating with the antigen solution.

Appendix 2: ELISA using antigen-coated wells (Modified from Dewey & Brasier, 1988)

Materials
- Multiwell polystyrene strips Microstrip, PS; strip holder frame; strip retainer (ICN Biomedicals Ltd).
- Phosphate buffered saline (PBS), pH 7.4

NaCl	8 g
KH_2PO_4	0.2 g
$Na_2HPO_4 . 12H_2O$	2.9 g
KCl	0.2 g

  Make up to 1 litre with distilled water.
- PBS–Tween — PBS with 0.05% Tween 20.
- Goat anti-mouse IgG + IgM–horseradish peroxidase conjugate (Sigma).
- Substrate buffer
50 ml 0.2 M sodium acetate, 1.95 ml 0.2 M citric acid, 100 ml distilled water.
- Substrate solution

Stock solution: $100\,mg\;ml^{-1}$ 3,3′,5,5′-tetramethylbenzidine in dimethyl sulphoxide.

Store at 4°C.

Dilute $150\,\mu l$ stock solution in $15\,ml$ substrate buffer and add $5\,\mu l$ H_2O_2 (30%) immediately before use.

- Antigenic material (see Appendix 1).

Method
- Coat surface of polystyrene wells with antigenic material diluted in PBS, $50\,\mu l$ per well. Incubate overnight at 4°C.
- Invert plate and flick out residual antigen solution.
- Wash four times with PBS−Tween, flicking out residual solution each time.
- Wash once with PBS.
- Rinse briefly with distilled water. Dry immediately by banging the plate, reverse side down, on absorbent paper. Finish drying by leaving in laminar airflow hood for 10 min at room temperature.
- Store sealed at 4°C until ready for use.
- Add $50\,\mu l$ undiluted CSF to each well (or if testing polyclonal antiserum dilute in PBS−Tween, generally 1 in 1000); leave at room temperature for 2 h (time may be adjusted).
- Wash four times with PBS−Tween.
- Add $50\,\mu l$ goat anti-mouse IgG + IgM−horseradish peroxidase conjugate diluted 1 in 200* in PBS−Tween, for 30 min at room temperature.
- Wash four times with PBS−Tween.
- Add substrate solution and leave for 30 min to develop.
- Add $50\,\mu l$ $2\,M$ H_2SO_4 to stop the reaction.
- Read absorbance at 450 nm with an ELISA plate reader.

Appendix 3: Dot-blot and dip-stick immunodiagnostic assays

Materials
- PVDF membrane (Immobilon P from Millipore). Wear rubber gloves and use clean equipment when cutting membrane. Use a soft pencil to mark membrane into squares for dot-blot assays or $30 \times 5\,mm$ strips for dip-stick assays.
- Tris buffered saline (TBS)
    20 mM Tris-HCl buffer, pH 8.2, containing 0.9% NaCl. (pH is critical.)
- 0.1% BSA.
- 20 mM sodium azide.
- TBS wash: TBS; 0.1% BSA; 0.05% Tween**.
- Blocking solution: TBS; 5% BSA; 0.05% Tween**.
- Antibody−gold conjugate: Janssen Auroprobe BL plus goat anti-

---

* Dilution may be adjusted; specificity of detector−antibody conjugate should reflect requirements of screening procedure, e.g. use anti-IgG specific conjugate if selecting only IgG MAbs.

** Tween 20 (0.05%) may or may not reduce non-specific binding; Janssen recommend that it is not used with their products and proteins.

mouse IgG + IgM (Amersham; product no. 30.695.43.).
- Silver enhancement kit: IntenSE BL (Amersham).

Method
- Wet the membrane by immersing it in 100% methanol for 1–2 min.
- Rinse the membrane briefly in distilled water and shake off surface water.
- Spot 3 µl of the antigen solution onto membrane (while it is still wet) or place strips of prewetted membrane (dip-sticks) in antigen solution in Eppendorf tubes and leave overnight at 4°C.
- Allow the antigens to air dry onto membrane at room temperature for 15 min.
- Rewet the membrane in methanol and distilled water as in the first two steps above.
- Incubate the membrane on a shaker in blocking solution for 1 h.
- Wash three times for 5 min in TBS wash.
- Incubate in primary antibody solution for 2 h (rocking).
- Wash three times for 5 min in TBS wash.
- Incubate on a shaker with antibody–gold conjugate (1/100) for 1 h.
- Wash twice in TBS wash.
- Wash in distilled water for 1 min.
- Incubate on a shaker with silver enhancer:initiator (1:1 v/v) for 10–20 min.
- Wash in an excess of distilled water.
- Dry between filter paper.

Appendix 4: Immunofluorescence assay

The easiest way to screen several CSF against a specific fungus is to grow the fungus on multiwell test well slides, as follows.

Materials
- Teflon-coated multiwell test slides (ICN Biomedicals Ltd).
- Sterile Petri dishes.
- Nutrient agar medium.
- Fluorescein isothiocyanate (FITC)–antibody conjugate.

Method
- Sterilize several multiwell slides by wrapping them individually in foil and autoclaving at 121°C for 15 min.
- Place sterile slides aseptically in sterile Petri dishes (glass or plastic) with Teflon side uppermost.
- Pour in nutrient agar medium until the slide is covered to a depth of 5 mm. Allow to set.
- Cut 5 mm wells through the agar over the clear areas in the slide using a cork borer (sterilized by dipping in alcohol and flaming) and remove the agar plugs with a sterile needle.
- Inoculate with spores or mycelial fragments around the well areas as indicated in Fig. 3.1.
- Incubate at an appropriate temperature until the fungus has grown over the surface of the agar and down and across the clear areas of glass at the base of the wells.
- Carefully remove the agar mat, take the slide out of the Petri dish, clean the reverse side and allow to air dry for at least 1 h.

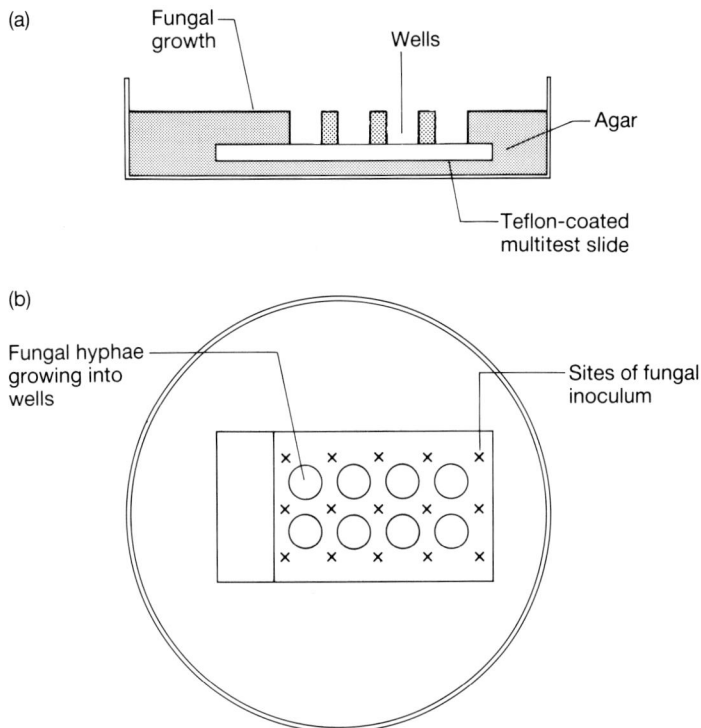

Fig. 3.1 Fungal growth on multiwell test slides for immunofluorescence screening of hybridoma cell culture supernatant fluids. (a) Side view. (b) View from upper surface.

- Fix fungal colonies by successively immersing the slide in:
    ethanol:chloroform:formalin (6:3:1 v/v/v) for 3 min.
    95% methanol for 4 min.
    distilled water for 30 s.
- Heating is not recommended as a method of fixation.
- Air dry completely at room temperature.
- Place the slide on moistened filter paper in a clean Petri dish. Pipette 10–20 µl of the CSF to be tested into the wells. Use one well for cell culture medium as a negative control and another with the polyclonal antiserum diluted approximately 1 in 40 as a positive control.
- Incubate the slide at room temperature for 30 min.
- Suck off antibody solutions gently and wash the slide three times for 5 min with PBS.
- Remove as much moisture as possible, drying the areas around the wells but not the wells themselves.
- Add 10–20 µl FITC conjugate diluted in PBS and incubate at room temperature for 30 min.
- Wash the slide gently three times for 15 min in PBS and dry.

- Add mounting medium (Citi-fluor recommended, Citi-flour Ltd). Cover all the wells with one large coverslip.
- Examine under u.v. light for fluorescence. Check the positive (polyclonal antiserum) and negative controls first.

4 Assay methods for mycotoxins using monoclonal antibodies

A.A.G. CANDLISH, W.H. STIMSON &
J.E. SMITH

Introduction

Mycotoxins are secondary metabolites produced by filamentous fungi growing on agricultural crops, commodities, food and feed products (Smith & Moss, 1985). They are a diverse range of chemically distinct compounds with molecular weights of 150–500, which are produced especially on crops subjected to adverse conditions such as environmental stress induced by drought or temperature, mechanical or natural damage, microbial disease or excessive use of agrochemicals. Once harvested, fungal growth can occur during storage of the crop and may result in very high levels of toxins. When affected food or feed products are consumed, mycotoxins at low levels (μg kg^{-1} material) may induce carcinogenic, oestrogenic, mutagenic or teratogenic effects. However, when animals consume large amounts of mycotoxins or ingest low levels over a long and continuous period, three disease categories may be defined (Pier et al., 1980).

1 Acute primary mycotoxicoses: may occur when milligrams of mycotoxins are consumed. Examples include hepatitis, haemorrhaging, nephritis, necrosis or death.

2 Chronic primary mycotoxicoses: results in reduced productivity rates.

3 Secondary mycotoxin diseases: occur when low levels of mycotoxins are consumed over prolonged periods of time, and are a result of impairment to the native and acquired immunity to infectious diseases.

The inherent toxicity of mycotoxins and their involvement in the aetiology of certain human and animal diseases makes their detection in the environment essential. Legal limits on the maximum levels tolerated in foods and feeds have been set in various countries (Schueller et al., 1982), and to assess levels of contamination suitable detection methods have had to be developed.

Traditionally detection methods involved the sampling, purification and isolation of individual mycotoxins using techniques such as thin layer chromatography (TLC), high performance liquid chromatography (HPLC) and mini-column chromatography (Egan *et al.*, 1982). Recently there has been an increasing interest in developing simpler and more specific methods for detecting mycotoxins using immunoassay procedures (Chu, 1984; Candlish, 1991). Research has mostly involved the development of techniques such as radioimmunoassay (RIA), enzyme-linked immunosorbent assays (ELISA) and immunoaffinity column chromatography. Immunoassay methods have benefited further from the development of monoclonal antibodies (MAbs) using hybridoma technology (Kohler & Milstein, 1975). A range of MAbs to mycotoxins such as aflatoxins (AFs), ochratoxin A (OTA) and T-2 toxin (Smith *et al.*, 1986) have already been developed. This chapter describes the development of MAb-based immunoassays for AFs and OTA.

Monoclonal antibody development

Haptenization The first stage in the development of MAbs to a mycotoxin is the conjugation of the toxin molecule to a carrier protein. The low molecular weight, haptenic nature of mycotoxins means that they will not induce an immune response when injected directly into animals. Antibodies can only be raised against the mycotoxin molecule after conjugation to a carrier protein, which makes the mycotoxin antigenic. Methods used for the conjugation procedure vary depending on the toxin involved. OTA has a free carboxylic acid which may be coupled directly to a protein using carbodiimide (Chu *et al.*, 1976). AFs require chemical activation as no reactive groups are present on the AF molecule. This can be achieved by the reaction of the ketone group on the AF molecule with O-carboxymethyloxime (Chu *et al.*, 1977). The oxime derivative formed can then be coupled to proteins using a procedure similar to the carbodiimide method for OTA (Chu & Ueno, 1977).

For each mycotoxin two different carrier proteins are used to develop two different mycotoxin−protein conjugate preparations. One conjugate is used for immunization, whereas the other (using a different protein carrier) is used to detect the antibodies raised to the first. In this way any antibody detected will only be directed against the hapten. In our experience, keyhole limpet haemocyanin (KLH) is the most useful protein for immunization and bovine serum albumin (BSA) is the best protein carrier for testing for the presence of specific MAbs.

The conjugate, once formed, must be purified by dialysis to remove unconjugated mycotoxins and excess carbodiimides. The amount of toxin coupled to the protein molecule can be assessed by absorbance at a specific wavelength: 362 nm for AFB1; and

333 nm for OTA. Molar ratios of mycotoxin to protein of 20:1 for BSA and 450:1 for KLH give good results.

Immunization   The immunization of animals follows the usual procedures (Table 4.1). Female BALB/c × New Zealand Black F.1. hybrid mice, 8–12 weeks old, are injected intraperitoneally with 100–320 μg of mycotoxin–protein conjugate per injection. Immunization is enhanced by the addition of an equal volume of Freund's adjuvant to the conjugate for each injection, complete adjuvant (CFA) for the primary injection and incomplete adjuvant (IFA) for subsequent immunizations. The booster injection contains no adjuvant and consists of conjugate diluted in saline.

Fusion techniques   Techniques used in the fusion of murine cells to myeloma cells have been described in detail elsewhere (Campbell, 1984; Goding, 1986; Torrance, Dewey, this volume). However, in our experience there are some points of importance in the development of successful hybridomas secreting MAbs to mycotoxins and we have given details of these in Appendix 1.

Screening cell lines   A sensitive indirect ELISA has been developed to detect MAbs which are specific for mycotoxins. This assay format allows rapid detection of many hundreds of hybridoma cell lines secreting MAbs against mycotoxins to levels as low as 20 ng ml^{-1} of MAb in cell culture supernatant fluid (CSF) (media in which hybridoma cells are grown). The procedure used to select anti-AFB1 MAbs is given in Appendix 2.

Large-scale antibody production and purification

Production of large quantities of MAbs against mycotoxins can be achieved by *in vitro* or *in vivo* methods.

In vitro methods have not yet been applied to the large-scale production of anti-mycotoxin MAbs, but as demand increases for large and consistent quantities of commercially available MAbs against mycotoxins, particularly AFs, then these methods

Table 4.1 Immunization protocol to produce an immune response to a mycotoxin

Immunization	Route	Quantity (μg)	Day of injection
Primary	i.p.	100–320	1
Secondary	i.p.	100–320	14
Tertiary	i.p.	100–320	28
Booster	i.p.	100–320	42

i.p. = intraperitoneal.

will be employed. Basically they involve the use of fermenter technology allowing the growth of hybridoma cells in flasks, roller bottles, airlift fermenters, hollow fibre chambers or encapsulation batch fermenters.

At present *in vivo* methods are almost entirely used for the production of MAbs. Hybridoma cells as ascitic tumours are grown in the peritoneal cavity of histocompatible BALB/c or F.1. hybrid mice. The animals are primed by the injection of $0.5-1.0$ ml of pristane into the peritoneal cavity and after $1-3$ weeks the hybridoma cells are injected by intraperitoneal injection at a concentration of 2×10^7 cells ml^{-1} in RPMI 1640 medium. These form an ascites tumour in $1-3$ weeks, the fluid from which on removal may contain $2-10$ mg ml^{-1} of antibody, and a volume of $0-10$ ml can be obtained per animal. However, this is an inherently variable technique: a group of 100 animals injected with the hybridoma cell line secreting MAb against AFB1 will produce as much as 1000 ml of ascites fluid on one occasion and virtually no fluid on another occasion under apparently identical conditions.

The MAb from CSF or ascites fluid may be purified by three methods, details of which are given in Appendix 3:
1 Ammonium sulphate ($[NH_4]_2SO_4$) fractionation.
2 Ion-exchange chromatography.
3 Protein A — Sepharose column chromatography.
The third method is especially useful when hybridoma cell lines do not induce good ascites tumours. The MAbs can be produced in CSF and purified by this method with $10-60$ mg of pure antibody per litre of fluid. Recovery levels of 100% can be expected if antibody is eluted at the acidic pH. However, Protein A does not bind well to mouse IgG$_1$ (Harlow & Lane, 1988).

Immunoassay methods

Three basic formats have been used in the development of detection methods for mycotoxins using MAbs:
1 ELISA.
2 RIA.
3 Affinity column chromatography.

ELISA   An indirect ELISA for screening MAb-producing cell lines specific for AFs is described in Appendix 2. However, for the detection of the mycotoxins in plant material competitive methods are required.

Two different competitive modes are described in Appendix 4: (i) indirect competitive ELISA for OTA (Candlish *et al.*, 1988 a); and (ii) direct competitive ELISA for AFB1 (Candlish *et al.*, 1985).

In the indirect competitive ELISA for OTA, the colour produced is inversely proportional to the concentration of the OTA in the sample. A typical standard dose−response curve has a full range reading at A_{450} of approximately 1.5 with 0 ng OTA ml^{-1}. A sensitivity of 0.5 ng ml^{-1} may be obtained with a working range up to 250 ng ml^{-1}. The 50% point of 0.75 (A_{450}) is reached with 12.8 ng ml^{-1} of OTA.

Samples of contaminated barley can be tested for their OTA content by this indirect competitive ELISA. Sample preparation is simple and rapid: chloroform extraction is followed by purification by silica gel column chromatography. The eluting solvent is evaporated and the dried OTA is redissolved in buffer and analysed by ELISA (Candlish, 1987). The chloroform extract can also be prepared by partitioning with 1 M $NaHCO_3$ buffer; the OTA partitions into the buffer, which is then tested by ELISA (Candlish et al., 1988 a). With these preparation methods levels as low as 5 µg OTA kg^{-1} of barley can be detected, with no interference from other components of the sample.

The direct competitive assay for AFB1 described in Appendix 4 has a sensitivity 0.2 ng ml^{-1} with a working range up to 30 ng ml^{-1} and 50% inhibition at 2.7 ng ml^{-1}. The coefficient of variation within an assay was 3.9−11.7%, and between assays 5.9−14.8%. The variation in readings across all the wells of a 96-well microtitre plate, with 0 ng ml^{-1} of competing AFB1, was 2.5%. The method is highly specific for AFB1 and other AF metabolites reacted only slightly (Table 4.2).

The ELISA can be used as a simple procedure for the routine determination of AFB1 in samples such as maize, peanut kernels and peanut butter. There is virtually no sample matrix interference with the performance of the assay even after minimal sample preparation because of the high affinity of the MAb employed.

Table 4.2 Cross-reactivity of different aflatoxin metabolites with direct competitive ELISA (Candlish et al., 1985)

Metabolites	Cross-reaction (%)*
AFB1	100.0
AFG1	14.3
AFB2	12.6
AFM1	7.3
AFM2	4.4
AFG2	1.2

* These values are calculated as the mass of AFB1 which would give 50% maximum absorbance as a percentage of the mass of the compound of interest giving the same reading ($n = 3$).

Essential features of sample preparation for ELISA are:
1 Blend sample with methanol:water (55:45 v/v)
2 Filter blended sample.
3 Dilute filtrate with 0.1 M Tris-HCl buffer.

Average recoveries of AFB1 from samples artificially contaminated with AFB1 at levels of $6-400 \mu g$ kg^{-1} were 90–112.5%. With laboratory samples contaminated by a fungus producing AFB1 there was a strong positive correlation ($r =$ 0.97) between results obtained by ELISA and TLC techniques (Candlish *et al.*, 1987).

RIA A MAb-based RIA has been developed for the specific and sensitive detection of OTA and is described in Appendix 5.

Using this assay the cross-reactivity of the MAb to OTA and the following analogues was assessed: OTB, OTA and L-β phenylalanine. Only OTA and OTB displaced the [^{14}C]OTA. The levels required to achieve 50% displacement were 0.5 ng for OTA and 2.5×10^5 ng for OTB. The RIA was sensitive to 0.075 ng of OTA per assay which corresponded to a minimum detection level of 0.2 ng of OTA g^{-1} of sample.

Samples were prepared by extraction of OTA with 0.5% phosphoric acid in chloroform partitioned into 1 M NaHCO$_3$ and applied to a Sep-Pak® C18 cartridge (Waters Associates). The OTA eluted from the cartridge was dried and applied to a Sep-Pak® silica gel cartridge. The eluted OTA was again dried, redissolved in buffer and applied to the RIA. This method recovered 70–80% OTA from 0.2–20 ng of OTA g^{-1} of sample. There was a strong correlation between the RIA and HPLC measurements for 10 naturally contaminated and two negative samples (Table 4.3).

Affinity column chromatography AFs can be concentrated, purified and detected using a MAb coupled to Sepharose® CL-4B (Pharmacia Ltd). This method, which is both rapid and sensitive, may also be used to clean up samples when HPLC or TLC analysis are required. It is fully described in Appendix 6. Recovery of total AFs by this method can be seen in Table 4.4.

Samples contaminated with AFs are extracted with methanol:water (60:40 v/v), blended, diluted to 30% methanol using water and filtered. Portions of the filtrate (10 ml) are then passed through the affinity columns and assayed for their AF content. Average recoveries of total AFs from maize and peanuts by this method were 94% and 83% respectively.

There was a good correlation between the immunoaffinity method of AF detection and established HPLC methods of analysis (Table 4.5). Furthermore, affinity columns can be used

Table 4.3 Comparison betweeen RIA and HPLC analysis of porcine kidney samples certified as negative and positive

Kidney sample	RIA (ng g^{-1})	HPLC (ng g^{-1})
Negative sample number		
1	<0.2*	<5.0*
2	<0.2*	<5.0*
Positive sample number		
7182	46	47.1
7197	36.4	35.7
7519	45.3	47.0
7537	85.9	88.7
7773	108.2	112.7
7783	50.2	51.5
7880	25.3	24.3
7894	21.7	21.7
7895	28.4	27.8
7904	4.3	4.0

*Detection limit = negative.

Table 4.4 Recovery of total aflatoxins from affinity columns

AF metabolite	Amount per column (ng)	Maize extract *	Peanut extract *
B1	25	90.5	87.1
B2	12.5	101.6	87.6
G1	25	95	91.1
G2	6.25	89.2	67.6

* Recovery expressed as percentages eluted from column.

Table 4.5 Comparison of affinity columns with HPLC for the determination of total aflatoxins in nuts, nut products and animal feedstuffs (parts per billion)

Sample	Affinity	HPLC
1	286	272
2	1	0
4	96	81
6	0	0
8	0	1
10	425	725
12	104	107
14	3.5	3
16	10.4	115
18	19	22
20	0	0
Wheat	0	0
Maize	0	0
328	2.5	3
602	3.9	43
619	8.5	10

for efficient sample clean-up prior to HPLC analysis. A range of affinity methods have now been developed for AFs (Groopman *et al.*, 1984; Candlish *et al.*, 1988 b).

Conclusion

Assay methods for the determination of mycotoxins based on immunochemical techniques using MAbs have been successfully developed. They are highly specific and sensitive when compared to traditional methods of detection and their simplicity and robustness allow them to be used routinely in the laboratory. Most attention has been focused on AFs and immunodiagnostic kits for AFs are now available commercially.

Techniques developed for the immunoassay of mycotoxins are not only useful to the plant pathologist for mycotoxin determination but also for the knowledge gained in developing immunoassay techniques for all type of haptens in plant materials. They should be readily applicable to the detection of pesticides and other low molecular weight metabolites such as growth regulators.

References

Campbell A.M. (1984) Monoclonal antibody technology. In Burdon R.H. & van Knippenberg P.H. (eds) *Laboratory Techniques in Biochemistry and Molecular Biology*, Vol. 13. Elsevier, Amsterdam.

Candlish A.A.G. (1987) *The development of monoclonal antibody-based enzyme immunoassays for ochratoxin A and aflatoxin B1*. PhD Thesis, University of Strathclyde.

Candlish A.A.G. (1991) The detection of mycotoxins in animal feeds using biological methods. In Smith J.E. & Henderson R.H. (eds) *Mycotoxins in Animal Foods*. CRC Press, Florida, in press.

Candlish A.A.G., Stimson W.H. & Smith J.E. (1985) A monoclonal antibody to aflatoxin B1: detection of the mycotoxin by enzyme immunoassay. *Letters in Applied Microbiology* **1**, 57–61.

Candlish A.A.G., Stimson W.H. & Smith J.E. (1987) The detection of aflatoxin B1 in peanut kernels, peanut butter and maize using a monoclonal antibody-based enzyme immunoassay. *Food Microbiology* **4**, 147–53.

Candlish A.A.G., Stimson W.H. & Smith J.E. (1988 a) Determination of ochratoxin A by monoclonal antibody-based enzyme immunoassay. *Journal of the Association of Official Analytical Chemists* **71**, 961–4.

Candlish A.A.G., Haynes C.H. & Stimson W.H. (1988 b) Detection and determination of aflatoxins using affinity chromatography. *International Journal of Food Science and Technology* **23**, 479–85.

Chu F.S. (1984) Immunoassays for analysis of mycotoxins. *Journal of Food Protection* **47**, 562–9.

Chu F.S. & Ueno I. (1977) Production of antibody against aflatoxin B1. *Applied and Environmental Microbiology* **33**, 1125–8.

Chu F.S., Chang F.C. & Hinsdill R.F. (1976) Production of antibody against ochratoxin A. *Applied and Environmental Microbiology* **31**, 831–5.

Chu F.S., Hsia M-T.S. & Jun P.S. (1977) Preparation and characterisation of aflatoxin B1-1-(*O*-carboxymethyl) oxime. *Journal of the Association of Official Analytical Chemists* **60**, 791–4.

Egan H., Stoloff L., O'Neill I.K., Scott P., Cartegnaro M. & Bartsch H. (eds) (1982) *Environmental Carcinogens: Selected Methods of Analysis*. Vol. 5. *Some Mycotoxins*. International Agency for Research on Cancer Scientific Publications Number 44, Lyon.

Goding (1986) *Monoclonal Antibodies: Principles and Practice*, 2nd edn. Academic Press, San Diego.

Groopman J.D., Trudel L.J., Donahue P.R., Marshak-Ruthstein A. & Wogan G.N. (1984) High-affinity monoclonal antibodies for aflatoxins and their application to solid-phase immunoassays. *Proceedings of the National Academy of Science, USA* **81**, 7728–31.

Harlow E. & Lane D. (1988) *Antibodies: A Laboratory Manual*. Cold Spring Harbor Laboratory, Cold Spring Harbor, New York.

Kohler G. & Milstein C. (1975) Continuous cultures of fused cells secreting antibody of predefined specificity. *Nature* **256**, 495–7.

Kurstak E. (1985) Progress in enzyme immunoassays: production of reagents, experimental design and interpretation. *Bulletin of the World Health Organization* **63**, 793–811.

Pier A.C., Richard J.L. & Cyzewski S.J. (1980) Implications of mycotoxins in animal disease. *Journal of the American Veterinary Medicine Association* **176**, 719–24.

Rousseau D.M., Candlish A.A.G., Slegers G.A., Petegham C.H.V., Stimson W.H. & Smith J.E. (1987) Detection of ochratoxin A in porcine kidneys by a monoclonal antibody-based radioimmunoassay. *Applied and Environmental Microbiology* **53**, 514–18.

Schueller P.L., Stoloff L. & van Egmond H.P. (1982) Limits and regulations. In Egan H., Stoloff L., O'Neill I.K., Scott P., Cartegnaro M. & Bartsch H. (eds) *Environmental Carcinocens: Selected Methods of Analysis* Vol. 5. *Some Mycotoxins*, pp. 107–116. International Agency for Research on Cancer Scientific Publications Number 44, Lyon.

Smith J.E. & Moss M.O. (1985) *Mycotoxins: Formation, Analysis and Significance*. John Wiley and Sons, Chichester.

Smith J.E., Candlish A.A.G., Stimson W.H. & Goodbrand I. (1986) Monoclonal antibodies. Their role in detecting and quantifying mycotoxins. In *Biotechnology in the Food Industry*. pp. 73–88. Online Publications, Pinner.

Appendix 1: Important points concerning cell fusions for producing monoclonal antibodies to mycotoxins

- Cell lines should be grown in bicarbonate buffered RPMI 1640 medium supplemented with 10% FCS and 2 mM glutamine.
- The source of lymphocytes must be the spleens of immunized mice.
- Give a booster immunization to the mice 2–4 days prior to fusion.
- Fuse the lymphocytes with myeloma cells P3 X 63-Ag8.653.
- Myeloma cells should only be used when growing in the exponential phase of growth.
- Select the hybridoma cells from the myeloma cells using HAT, which is selective against the myeloma cells but has no effect on hybridoma cells.
- Fuse the myeloma cells with lymphocytes using 1 ml of 46% (w/v) polyethylene glycol (mol. wt 1550) added over 3–5 min at 37°C.
- Hybridoma cells should be screened for specific MAb activity after 10–14 days by indirect ELISA. Screening of antibody should be continued at weekly intervals as the hybridoma cells grow.
- Clone the hybridoma cells by limiting dilution 2–3 months after fusion.

- Hybridoma cells, both cloned and uncloned, must be stored in liquid nitrogen as soon as possible, in media containing 20% dimethyl sulphoxide, 30% RPMI 1640 and 50% (v/v) FCS.

Appendix 2: Screening cell lines for monoclonal antibodies specific for aflatoxin B1

- Coat 96-well microtitre plates by adding $10\,\mu g\,ml^{-1}$ of AFB1−BSA conjugate in $0.02\,M$ Tris-HCl buffer, pH 9.2, containing $0.15\,M$ NaCl (coating buffer), $200\,\mu l$ per well, incubating plates at 37°C for 2 h.
- Wash three times with $0.2\,M$ Tris-HCl buffer, pH 7.4, containing $0.15\,M$ NaCl and 0.05% (v/v) Tween 20 (wash buffer).
- Block the wells with 1% BSA in coating buffer by incubating the solution in the wells for 30 min at 37°C.
- Wash three times with wash buffer.
- Add $100\,\mu l$ per well of CSF to the coated wells. Incubate for 1 h at 37°C.
- Wash five times with wash buffer.
- Add $100\,\mu l$ per well of anti-mouse IgG horseradish peroxidase conjugate at 1:1000 dilution in wash buffer. Incubate for 1 h at 37°C.
- Wash five times with wash buffer.
- Add $100\,\mu l$ per well of $60\,\mu g\,ml^{-1}$ of tetramethylbenzidine (TMB) in $0.1\,M$ sodium acetate/citric acid buffer, pH 5.0, containing 0.03% H_2O_2. Incubate for 30 min at room temperature.
- Stop the reaction with $2\,M$ H_2SO_4 and measure the absorbance at 450 nm.

On completion of the ELISA any hybridoma cell line that secretes specific MAb to AFB1 will give a positive coloured product. If the hybridoma secretes a non-specific MAb, or none at all, then no antibody binding will occur to the AFB1 attached to the solid-phase microtitre wells and hence no colour will result.

Only MAbs specific for AFB1 and not the carrier protein are selected in this ELISA, as the protein used as carrier for immunization was KLH. No cross-reactivity between these two carrier proteins has been observed.

Appendix 3: Purification of monoclonal antibodies from cell culture supernatant fluids or ascites

Ammonium sulphate fractionation

Add an equal volume of saturated ammonium sulphate to the ascitic fluid and leave for 1 h at 4°C (to precipitate the MAb). Precipitated proteins are collected by centrifugation ($10\,000\,\boldsymbol{g}$ for 20 min) and the pellet washed twice with 50% saturated ammonium sulphate. The pellet is then resuspended in a suitable buffer and its concentration adjusted to $10\,mg\,ml^{-1}$.

The ammonium sulphate fraction can be further purified by ion-exchange chromatography.

Ion-exchange chromatography

The MAb solution is dialysed overnight against $10\,mM$ phosphate buffer, pH 7.5, and then added to a 'S' Sepharose® fast-flow cation exchange column (Pharmacia) to which the antibody binds. The column is washed with $10\,mM$ phosphate buffer, pH 7.5. The bound protein containing the specific MAb is then eluted with 10 mM phosphate buffer, pH 7.4, containing 50 mM NaCl. Elution of protein is monitored by absorbance at 280 nm and specific antibody activity estimated by indirect ELISA.

Protein A —
Sepharose® column
chromatography

This has also been used to purify the specific MAb to AF from large volumes of CSF as follows.
● Swell 1.5 g protein A-Sepharose® CL-4B (Pharmacia) in 0.05 M Tris-HCl buffer, pH 8.6, containing 0.15 M NaCl and 0.1% sodium azide. Pack resin to a bed volume of 5–6 ml in a suitable column.
● Centrifuge CSF at 200 g for 10 min. Adjust the pH of the supernatant to 8.6 with 1 M NaOH.
● Apply 1 litre of the supernatant to the protein A column at a flow rate of 50 ml per hour.
● Elute MAb from the protein A column by stepwise pH elution with the following buffers at a flow rate of 50 ml per hour for 1 h per buffer:
   0.05 M phosphate, pH 7.0, containing 0.15 M NaCl;
   0.05 M citrate, pH 5.5, containing 0.15 M NaCl;
   0.05 M acetate, pH 4.3, containing 0.15 M NaCl;
   0.05 M glycine-HCl, pH 2.3, containing 0.15 M NaCl.
● Monitor specific MAb activity by indirect ELISA in each of the different fractions.
● Pool fractions containing the antibody and dialyse against the appropriate buffer. For the MAb against AFB1 fractions with pH values of 5.5 and 4.3 contain the appropriate activity.
● Regenerate the column by washing and equilibrating with 0.05 M Tris-HCl buffer, pH 8.6, containing 0.15 M NaCl and 0.1% sodium azide.

Appendix 4: ELISAs for mycotoxins

Indirect competitive
assay for OTA

● Coat the wells of a microtitre plate with OTA–BSA conjugate at a concentration of 5 µg ml^{-1} by adding 100 µl per well diluted in 0.02 M Tris-HCl buffer, pH 9.2. Incubate for 2 h at 37°C.
● The untreated ascites fluid from a hybridoma cell line secreting anti-OTA MAb is diluted 1:40 000 (v/v) in 0.1 M Tris-HCl buffer, pH 8.5, and added to the wells (100 µl per well) together with standard or sample OTA (50 µl).
● The plate is incubated for 1 h at 37°C, during which time the MAb interacts and competes equally for free and solid-phase OTA; the more OTA present in the standard or sample, the less antibody will bind to the solid-phase toxin.
● Wash to remove unbound antibody, i.e. antibody bound to free OTA.
● An anti-mouse IgG–horseradish peroxidase conjugate (diluted 1:2000 (v/v) in 0.9% saline containing 25% normal sheep serum) is added to monitor the amount of anti-OTA MAb bound. Incubate at 37°C for 1 h.
● After incubation the plate is washed and the enzyme substrate (TMB) added as in Appendix 2. The colour produced is inversely proportional to the concentration of OTA in the standard or sample.

Direct competitive assay for AFB1

Preparation of MAB–horseradish peroxidase conjugate by the periodate method (Kurstak, 1985)

- To $10\,mg\,ml^{-1}$ horse radish peroxidase in $1\,ml$ of $0.1\,M$ $NaHCO_3$ solution add $0.05\,ml$ $0.05\,M$ sodium periodate.
- Mix for $2\,h$ in the dark to activate the enzyme and add it to $1.5\,ml$ of the MAb in $0.1\,M$ $NaHCO_3$ solution, at a concentration of $10\,mg\,ml^{-1}$.
- The enzyme and MAb mixture are added to $0.33\,g$ of Sephadex® G-25 (Pharmacia) and placed in a sealed Pasteur pipette. The beads swell immediately, concentrating the protein solution and absorbing the excess periodate.
- Incubate in complete darkness for $2-3\,h$ and elute the conjugate from the Sephadex® by the addition of $0.1\,M$ carbonate buffer, pH 9.3.
- The conjugate is stabilized by adding $0.5\,ml$ of $0.1\,M$ diethanolamine and stored as aliquots at $-20°C$ until required.

Direct competitive assay (Candlish et al., 1987)

The MAb–peroxidase conjugate is used in the direct competitive ELISA as follows.
- Coat the wells of 96-well microtitre plates with AFB1–BSA ($5\,\mu g$ ml^{-1}, $200\,\mu l$ per well) as for indirect ELISA for MAb detection.
- Add $100\,\mu l$ per well of MAb–peroxidase conjugate at a suitable dilution (approx. 1:3000, v/v) in $0.15\,M$ NaCl containing 25% (v/v) normal sheep serum.
- Immediately add $50\,\mu l$ per well of standard or sample AFB1 in $0.1\,M$ Tris-HCl buffer, pH 8.5.
- Incubate for $1\,h$ at $37°C$.
- Wash five times with wash buffer.
- Add $150\,\mu l$ per well of TMB substrate as in Appendix 2 and incubate for $30\,min$ at room temperature.
- Stop the reaction with $50\,\mu l$ per well of $2\,M$ H_2SO_4 and measure the absorbance at $450\,nm$

The reading is plotted against the log of AFB1 concentration in ng ml^{-1} and estimation of AFB1 in samples in assessed by reference to the standard curve.

Appendix 5: Radioimmunoassay for OTA (Rousseau *et al.*, 1987)

The ascites fluid containing the MAb is used in a solid-phase IgG RIA on protein A-Sepharose® CL-4B (Pharmacia) with [^{14}C]OTA as tracer.
- The antibody is coupled to Sepharose® by incubating $20\,\mu l$ of a 1:100 (v/v) dilution of crude ascites fluid with $8\,mg$ (dry weight) preswollen protein A-Sepharose® CL-4B (Pharmacia) for $1.5\,h$ at $37°C$.
- Wash with $0.1\,M$ sodium phosphate buffer (NaPB), pH 7.4.
- Add $100\,\mu l$ of standard or sample solutions of OTA in buffer and $100\,\mu l$ of [^{14}C]OTA (8000 c.p.m.) in $0.1\,M$ NaPB, pH 7.4, to the MAb-coupled protein A-Sepharose®.
- Incubate for $4\,h$ at $37°C$ with mixing.
- Collect the supernatant by centrifugation at $710\,g$ for $30\,s$.
- The gel is washed twice with $0.5\,ml$ of $0.1\,M$ NaPB, pH 7.4, and centrifuged at $710\,g$ for $30\,s$.
- The supernatants are pooled and c.p.m. measured.

Appendix 6: Affinity chromatography

- Suspend freeze-dried CNBr-activated Sepharose® 4B (10 g) (Pharmacia) in 300 ml of 1 mM HCl over a sintered glass funnel.
- Wash the gel with 50 ml of 0.2 M NaHCO$_3$ buffer, pH 8.7, containing 0.5 M NaCl.
- Add to purified MAb solution (Appendix 3) (0.25 mg ml^{-1}) in 70 ml of the same buffer.
- Mix the MAb and Sepharose® for 2 h at room temperature and filter the suspension through a sintered glass funnel.
- Block any active groups remaining on the Sepharose® by washing and mixing with 0.1 M Tris-HCl buffer, pH 8.0, for 1–2 h at room temperature.
- Filter through a sintered glass funnel and wash with three to five cycles of two buffers of alternating pH. The first buffer is 0.1 M Tris-HCl, pH 8.0, containing 0.5 M NaCl and the second buffer is 0.1 M sodium acetate/acetic acid, pH 4.0, containing 0.5 M NaCl.
- The MAb-bound Sepharose® is washed and transferred to 70 ml 0.1 M PBS, pH 7.4, containing 0.02% thimerasol, to give a total volume of 105 ml. This may be stored at 4°C for up to 12 months without loss of activity.

This affinity matrix can then be used in an immunodiagnostic test for total AFs. Affinity columns are prepared as follows.

- Add 0.5 ml of MAb-coupled Sepharose to 55 mm × 5.5 mm columns.
- Pass 10 ml of a 30% (v/v) solution of methanol in water containing standard concentrations of AFs through the column at 5 ml per min.
- Wash twice with 10 ml of distilled water.
- Elute the AFs bound to the MAb with 1.0 ml of methanol.
- The AF content in the methanol is assessed by measuring fluorescence in Florisil® under u.v. light or by HPLC analysis after dilution with 4 ml of acetonitrile:water (60:40, v/v).

5 Techniques for detecting and identifying plant pathogenic bacteria

D.E. STEAD

Introduction

Bacterial diseases of plants are most often diagnosed on the basis of typical symptoms and subsequent isolation and characterization of the pathogen by means of classical morphological, physiological and nutritional tests. These methods are well described in popular books on diagnosis (Fahey & Persley, 1983; Lelliott & Stead, 1987; Schaad, 1988). For most diseases presumptive diagnoses made by such methods may take a minimum of 1 week. Should confirmation be necessary a host test may take a further week or more. Thus, for some slow-growing pathogens, for example *Clavibacter michiganensis* subsp. *sepedonicus*, diagnosis may take more than 1 month. Such lengthy diagnoses are often unacceptable. Similar problems also apply to detection of low numbers of pathogens present in, for example, asymptomatic tissues, seed samples, or epiphytic populations.

Isolation may be hampered by problems with selection of suitable infected asymptomatic tissues, overgrowth by other endophytic, epiphytic or saprophytic bacteria, or of being unable to recover the pathogen when using selective media. In addition, most of the classical nutritional tests for characterization based upon the ability of a pathogen to utilize specific compounds rarely give clear-cut identification since ability to utilize many compounds can vary within strains of a given taxon. Such methods are often inaccurate, slow and labour-intensive and more rapid and accurate methods are required. In this context the word 'rapid' implies less time taken by a technician to complete a test as well as less time taken from receipt of specimen to completion of diagnosis. This chapter reviews some techniques which allow more rapid and accurate diagnosis and more sensitive detection than the classical methods.

There are two major strategies.

1 Produce a profile of a suitable series of cellular components or reactions which can be obtained quickly and compared against a library of profiles.

2 Produce specific reagents which react only with the target bacterium in a rapid, easy-to-use test.

Profiles of cellular components or reactions usually require isolation and purification of the test organism. Rapid analysis then usually allows diagnosis to be made within a few days or even hours of obtaining a pure culture. Cellular components which lend themselves well to such analyses include lipids, proteins, plasmids and restricted DNA fragments.

The use of specific reagents allows detection directly within plant tissues, thus removing the necessity for isolation. These methods may allow detection and diagnosis within hours or even less time.

The major techniques are listed in Table 5.1 and their value in identification of cultures and detection of low numbers of bacteria in plant tissue is indicated. Nucleic acid technology is discussed by Vivian (this volume). This chapter reviews the following techniques and gives currently acceptable protocols for each.

1 Serological methods including immunofluorescence (IF), enzyme-linked immunosorbent assay (ELISA) and immuno-diffusion.

2 Nutritional profiles using commercially available kits.

3 Fatty acid profiles using gas chromatography.

Table 5.1 List of techniques and their uses

Technique	Duration	Identifying cultures	Detection in — Diseased plants	Detection in — Subclinical infections	Large scale
Immunodiffusion	1−2 days	++	+	−	−
Slide agglutination	5−10 min	+	+	−	+
Latex agglutination	5−60 min	++	++	+	++
Protein A agglutination	10−60 min	++	++	+	++
IF	2−3 h	++	++	+	++
Immunosorbent-IF	3−4 h	++	++	+	++
ATA ELISA	4−12 h	++	++	+	++
Dot-immunobinding	4−12 h	++	++	+	++
IF colony staining	1−3 days	++	++	++	+
Immunoisolation	2−3 days	−	++	++	+
Nutritional profile kits	4 h−2 days	++	−	−	+
Fatty acid profiles	3−4 h	++	+	−	++
Protein profiles	6 h−2 days	++	−	−	+

* Times given are approximate and do not include time for isolation and purification where necessary.
++ = useful with some limitations; + = very limited application; − = not applicable.

4 Protein profiles using polyacrylamide gel electrophoresis.

Protocols for most of the techniques listed in Table 5.1 may also be found in Klement *et al.* (1990) and many are also discussed by Austin & Priest (1986).

Serological techniques

Some of the advantages and disadvantages of using serological tests for diagnosis of plant pathogenic bacteria have already been discussed in Lelliott & Stead (1987). This section aims to review some of the newer and some of the most useful established techniques for their use both in detection and diagnosis. Other chapters in this volume give protocols for serological assays for viruses (Torrance), fungi (Dewey) and mycoplasma-like organisms (Clark), some of which are also discussed here with reference to bacteria. A wide range of serological assay protocols are also described by Schaad *et al.* (1990).

Of all the techniques discussed in this chapter serological techniques offer the best potential methods for detection of bacteria directly in plant material, for example in seed samples, in subclinical infections, in soil and plant debris or for detection of epiphytic bacteria.

The success of any serological technique depends upon two main criteria: (i) the specificity and sensitivity of the antiserum; and (ii) the methods used to amplify and observe the serological reaction. There are two major types of preparations of antibodies: (i) polyclonal antisera produced in the animal in response to the antigen and containing antibodies specific for more than more than one epitope; and (ii) monoclonal antibodies (MAbs) prepared by cell culture techniques and produced by a hybridoma (a single lymphocyte cell hybridized with a tumorigenic cell) specific for a single epitope. MAbs have the advantage over most polyclonal antisera of being specific for particular epitopes but care must be exercised when choosing the epitopes to be detected (Torrance, this volume). This may be because the epitopes to which they are specific are more scarce on the cell surface and therefore it is more difficult to detect the antigenic reaction without recourse to some form of amplification.

Suitable schedules for producing polyclonal antisera are given by Lelliott & Stead (1987). MAbs have now been produced for a wide range of pathogenic bacteria. The basic methods for their production are reviewed by Macario & de Macario (1988) but see also Torrance (this volume). Table 5.2 lists examples where MAbs have been raised to plant pathogenic bacteria.

The recent emphasis on the use of MAbs for diagnosing plant pathogenic and other bacteria has highlighted the importance

Table 5.2 Bacterial plant pathogens to which monoclonal antibodies have been raised

Species of bacterium	Reference
Clavibacter michiganensis subsp. *sepedonicus*	De Boer & Wieczorek (1984); De Boer & McNaughton (1986); Magee *et al.* (1986); De Boer *et al.* (1988)
Erwinia carotovora subsp. *atroseptica*	De Boer & McNaughton (1987)
Xanthomonas campestris pv. *dieffenbachiae*	Bonner *et al.* (1987)
Xanthomonas campestris pv. *oryzae*	Benedict *et al.* (1989)
Xanthomonas campestris pv. *oryzicola*	Benedict *et al.* (1989)
Xanthomonas campestris pv. *campestris*	Yuen *et al.* (1987); Alvarez *et al.* (1985)
Xanthomonas campestris pv. *citri*	Alvarez *et al.* (1987); Benedict *et al.* (1985)

of selecting the antigen to be detected. It is now possible to purify chemical components unique to specific taxa (lipopolysaccharides, siderophore, cell wall peptidoglycans, soluble and other cellular proteins, enzymes and their isozymes) and produce antisera to them by the standard methods used to produce polyclonal antisera in rabbits. Affinity chromatography and cross-absorption techniques can also be of great use in increasing the specificity and sensitivity of polyclonal antisera.

Immunodiffusion and agglutination techniques

These techniques have been used for many years to identify plant pathogenic bacteria and will not be described in detail here. Nevertheless, it should not be forgotten that immunodiffusion (Weir, 1978; van Regenmortel, 1982; Lelliott & Stead, 1987) is still one of the best methods available for the identification of small sample numbers of cultures of plant pathogenic bacteria and that it is relatively rapid.

Slide agglutination gives a result in a few minutes but only when there are large numbers of cells present (about 10^9 cells ml^{-1}). It does however require relatively large quantities of antisera and it can only be used to identify bacteria which are serologically dissimilar to other closely related bacteria. We use the technique routinely to identify cultures of *Erwinia amylovora* and have found very few cross-reactions over many years of use. It is possible to apply slide agglutination to detect a pathogen directly in plant material. Antisera produced to whole cells are not suitable in slide agglutination to identify bacteria such as *Erwinia carotovora* subsp. *carotovora* or most pathovars of *Xanthomonas campestris* and *Pseudomonas syringae* since these contain many common antigens or, as with *E. carotovora* and *E. chrysanthemi*, many different serogroups are found.

Latex agglutination is less wasteful of antisera than slide agglutination and is more sensitive; it occurs even in the presence of soluble antigen. It can be a useful rapid screen when used directly on plant material for plus or minus detection when samples with negative results are assumed to be free of disease. Those with positive results may require further testing. Yet another novel agglutination technique uses the protein A on *Staphylococcus aureus* cells (Chirkov *et al.*, 1984).

Immunofluorescence (IF) techniques (Appendix 1)

These are perhaps the most widely used serological techniques, especially for detection of bacteria directly from plant material (Stead, 1987; Lelliott & Stead, 1987). The most commonly used fluorescent agent is fluorescein isothiocyanate (FITC) which is conjugated to anti-bacterial antibodies in direct tests, or to an anti-rabbit (or mouse) immunoglobulin in indirect tests. Indirect tests usually give clearer differentiation of positive and negative results. Additionally, provided that all antisera to target antigens are produced in the same animal species, a single conjugate purchased commercially such as anti-rabbit IgG−FITC can be used for all tests.

All IF tests must take into account the possibilities of antigen and antibody excess, and dilutions of both must be tested. When determining antiserum titre it is essential to standardize the bacterial population numbers. I use a 10^6 cell ml^{-1} suspension on the slides which usually provides adequate cell numbers with no problems of antigen excess. Titres should be determined for a wide range of strains of the target organism including the homologous strain to which the antiserum was produced. Titre is determined as the reciprocal of the highest dilution at which maximum fluorescence was observed. Then, a working dilution should be determined as the highest dilution which would give a clear positive result with all known strains, usually about two or three doubling dilutions below the homologous titre.

When using IF to identify bacterial cultures the cell concentration should be adjusted to 10^6 cells ml^{-1}. Immunofluorescence detects dead as well as live cells so a total count rather than a viable count is better for calculating population numbers. Detection of the target bacterium in plant tissues where the numbers cannot be adjusted should be performed by checking a series of 10-fold dilutions, e.g. duplicate sap samples of undiluted (neat) 1:10, 1:100 and 1:1000 dilutions with a buffer control in place of antibody to check non-specific binding of the FITC conjugate. These tests involve as many as 10 windows on a multispot slide. The 1:10 or 1:100 dilutions often give the best results. Neat preparations contain much plant debris, especially with starchy samples, and heat-fixed preparations can be easily washed off during washing.

Immunosorbent immuno-fluorescence

This is a modification of the direct IF test in which the microscope slide windows are coated with commercial nail varnish in acetone (33% v/v) before addition of antiserum diluted in carbonate buffer, pH 9.6, and then sample, followed by FITC conjugated directly to the antiserum produced against the target bacterium (van Vuurde & van Henten, 1983; Stead, 1987; van Vuurde, 1987).

Immuno-fluorescence colony staining

The test sample is mixed with molten agar at 40°C and a conventional pour plate prepared. After 12–24 h growth the plates are dried and then flooded with an appropriate dilution of FITC-labelled antibodies against the target bacterium. The plates are then checked at low magnification using an u.v. microscope for fluorescing colonies. This type of assay has been used successfully for a wide range of plant pathogenic bacteria (van Vuurde, 1987).

ELISA (Appendix 2)

There are many different types of ELISA. All use an enzyme-mediated colour change to indicate presence of an antigen (Torrance, this volume). In PTA (plate trapped antigen) ELISA the antigen is attached directly to a microtitre plate. In ATA (antibody trapped antigen) ELISA the antigen is trapped by a layer of antibodies coated on the surface of the microtitre plate. Unfortunately, in PTA ELISA if the antigen comprises whole bacterial cells they may be easily washed off the plates, so it is wise to try several different brands of microtitre plates with each antigen system used. If attachment of antigen is a problem plates can be coated with poly-L-lysine first, then, after addition of antigen, plates are centrifuged to bring antigen into close contact with the poly-L-lysine and the antigen is fixed to it with glutaraldehyde (Appendix 2).

Bacterial cells also produce soluble antigens. Detection of these presents no such problems and a typical ATA ELISA technique is given in Appendix 2. However, it is essential to block the plates with gelatin or skimmed milk to prevent non-specific binding to uncoated areas of the microtitre plate wells (Appendix 2).

When using ATA ELISA techniques if the anti-bacterial

immunoglobulins are not conjugated directly to enzyme it is essential to use antibodies prepared in two different hosts to prevent the anti-immunoglobulin−enzyme conjugate attaching directly to the trapping antibodies coated to the surface of the plates. If antibodies from two animal hosts are not available it is possible to use the protein A sandwich ELISA technique of Edwards & Cooper (1985) or the F(ab')$_2$ ELISA technique of Barbara & Clark (1982). When using MAbs in ATA ELISA it is common practice to first use a polyclonal antibody as the trapping antibody. Protocols for some of these procedures are also given in this book (Torrance; Dewey; Clark).

A number of different enzymes are used in ELISA; these include alkaline phosphatase, horseradish peroxidase and urease. Viable bacteria can produce enzymes which use the same substrates and it is possible that the substrate may be broken down by exogenous bacterial enzymes giving false positive results. Adequate controls must therefore be included and the most suitable enzyme used. Alkaline phosphatase is the most widely used for the detection of plant pathogenic bacteria.

Most publications on the use of ELISA for bacterial diseases refer to the specifity and sensitivity of the test using pure cultures. Relatively few publications investigate the potential of ELISA to detect bacteria in plant tissues. However, ELISA has been used successfully to detect latent ring rot of potato caused by *Clavibacter michiganensis* subsp. *sepedonicus* (De Boer & McCann, 1989).

There are now several commercial ELISA kits available for diagnosis of bacterial diseases. These have two potential limitations. They may not be as sensitive as other serological assays such as IF, and the antisera they use may have limited specificity. Therefore, they may be of limited value in the detection of subclinical infection and give false positive results. Such kits require full evaluation before routine use. A recently introduced MAb-based ELISA kit to detect bacterial blight of pea (*Pseudomonas syringae* pv. *pisi*) has been used to identify bacterial cultures.

Dot-immunobinding assays

This is a variation of ELISA in which nitrocellulose or nylon membranes are used as the inert substrate. In most cases the target antigen is bound directly to the membrane and then an indirect procedure is carried out comprising antibody to target bacterium (e.g. produced in rabbit) followed by anti-rabbit immunoglobulin conjugated to an enzyme (Barnett, 1986). In contrast to most ELISAs the enzyme substrate used forms an insoluble coloured product, so that a positive result appears as a coloured dot on the white nitrocellulose or nylon background.

Such assays may be simpler and more rapid than ELISA but since bacterial adsorption to the membrane may be poor, it may be necessary to introduce a more efficient trapping stage or use soluble antigen rather than whole cells. A protocol for dot-immunobinding assay is given in this volume (Torrance).

Immunoisolation In immunoisolation the high affinity of antibody for antigen is used to trap the target bacterium allowing all other bacteria to be washed away. The target bacterium is then released into a growth medium. In its simplest form a glass rod is first coated with nail varnish in acetone (33%, v/v) and then with antibodies. The coated rod is used to crush a sample of diseased tissue. After washing the glass rod, it is streaked over the surface of an agar plate (van Vuurde et al., 1986; van Vuurde, 1987; Stead et al., 1987).

Alternatively, wells of microtitre plates can be coated with antibody before washing and filling with a growth medium (van Vuurde, 1987), or comminuted plant tissues can be 'filtered' through a column of antibody-sensitized beads before selective desorption and culture of the 'purified' target bacteria on agar plates (van Vuurde et al., 1986; Ruissen et al., 1986; Stead et al., 1987).

Nutritional profiles using commercially available kits

These are based on the classical cultural methods for identification of bacteria but they tend to be in a form which allows much greater reproducibility of results with a wider range of compounds. There are a number of commercially available kits and most are produced for use with specific groups of bacteria, primarily those groups of medical importance. Results can be determined according to different parameters:

1 assimilation of the carbon source observed as growth;

2 fermentation of the carbon source under anaerobic conditions, observed as acid production in the presence of a pH indicator;

3 oxidation of the carbon source under aerobic conditions, observed as acid/alkali production in the presence of a pH indicator;

4 detection of specific enzyme activity within a few hours and observed as the release of a specific indicator.

Some systems are based on dehydrated substrates in individual wells in galleries. Test cultures are suspended in a suitable basal medium and added to each well. After incubation they are observed for growth, acid production or enzyme activity. These systems include Micro-ID from General Diagnostics, which is based on 15 enzyme reactions and claims to identify

30 different enteric bacteria in 4 h, and the API systems (API-bio Mérieux (UK) Ltd) which rely on a series of tests. The API 20 E is specifically designed for Enterobacteriaceae and API 20 NE for a fairly wide range of Gram negative oxidative bacteria. Determination of positive results allows a code to be produced for each strain; the code is then identified using reference cultures or by using a reference book or computerized database supplied by the manufacturers.

The kits are useful for identifying some plant pathogens. A series of codes for many *Erwinia* spp. has been determined by Mergaert *et al.* (1984). Unfortunately some of the available databases are based on results obtained from kits where samples are incubated at 37°C for 24 h, conditions which are not suitable even for some of the plant pathogenic Enterobacteriaceae. Thus, databases need to be compiled from results obtained at more appropriate cultural conditions such as incubation at 28°C for 48 h, which is adequate for most plant pathogens. These conditions must be adhered to for all tests.

Of more value, but costing more, are the API 50 series. We now regularly use three of them:

1 API 50 CH: five strips each containing 10 different carbohydrates;

2 API AO: five strips each containing 10 different organic acids;

3 API AA: five strips each containing 10 different nitrogenous compounds.

Thus, for each strain, a nutritional utilization profile of up to 150 compounds can be obtained. One well in each series serves as a control. Pure cultures of the test bacterium are adjusted to 10^4 colony-forming units (cfu) ml^{-1} in an appropriate basal medium which may or may not contain a pH indicator. The *Pseudomonas* assimilation medium produced by API gives excellent results with pseudomonads. I prefer to look for carbon assimilation rather than acid/alkali production and to assess growth in the aerobic zone of the well. However, if fermentation results are required, growth in the anaerobic zone should be recorded. With approximately 150 different carbon sources available the results are good enough for numerical taxonomic studies. However, the kits are fairly expensive and it may not be cost effective to use all three routinely. They are excellent for determining carbon sources of diagnostic value and for characterizing new or unusual pathogens. The API 50 CH (carbohydrates) is the most commonly used kit and may be sufficient on its own for accurate identification of some bacteria, which would prove more cost effective than preparing the necessary agar and broth media in the laboratory.

However, the systems described so far cannot be modified

and are rather inflexible. Some of the others are more flexible. Discs impregnated with substrates can be purchased separately and kits tailored to a specific purpose can be made up in micro-titre-type plates. Again, a specific number of the test bacteria are suspended in a suitable basal medium and added to each well. Some systems, e.g. A/S Rosco Rapid ID tests (Lab M), rely on high levels of enzyme activity and use a selection of carbo-hydrates and various enzyme substrates or rapid *in situ* detec-tion of aminopeptidases, esterases, lipases and glycosidases. Others, such as the Minitek system (BBL), rely on colour changes resulting from the activity of enzymes produced during 18–24h active growth. Some plant bacteriologists have de-veloped their own kits, e.g. 96-well microtitre plates for use with plant pathogenic pseudomonads (Hayward *et al.*, 1989). A very recent commercial introduction, the Biolog System (Biolog Inc.) is based on the utilization of 95 different compounds in the wells of a microtitre plate. Utilization is based on redox chem-istry; increase in respiration during utilization is accompanied by a colour change indicated by a tetrazolium salt. The results can be assessed manually or read on a plate reader linked to a computerized library of culture collection strains so that a match with a similar coefficient can be given. It is likely that this flexible system will become popular. The library contains a wide range of plant pathogenic, as well as other, bacteria in-cluding many pathovars of *Xanthomonas campestris* and *Pseudo-monas syringae*, but since nutritional profile systems rarely, if ever, allow accurate identification at an infraspecific level, it is difficult to see why the Biolog system should be any better.

There are many publications which assess and compare these kits for rapid identification of the specific groups of bacteria listed below.

1 Enterobacteriaeae: Brennan *et al.* (1974); Hansen *et al.* (1974); Nord *et al.* (1975); Aquino Dowell (1975); Shayegani *et al.* (1975); Hanson *et al.* (1978); Smith *et al.* (1981); Maes-troeni *et al.* (1984); Villagarcia (1985); Holmes & Humphry (1988).

2 *Erwinia* spp.: Mergaert *et al.* (1984).

3 Oxidative Gram negative bacteria: Wellstood-Nuesse (1979); Burdash *et al.* (1980).

4 Anaerobes: Hanson *et al.* (1979).

5 Coryneforms: Slifkin *et al.* (1986).

Computerization It is a well-documented fact that some strains of a pathogen will have a variable reaction with almost any substrate, so use of dichotomous keys based on positive/negative results for individual tests often leads to false identifications. Thus, any form of computerized identification should be based on probability methods. Lapage *et al.* (1970) suggest that a minimum of 30−40 tests is required to obtain a high level of confidence in positive identification of all strains including aberrant strains. The use of computerized probability methods for identification of bacteria using such tests is discussed by Bascomb *et al.* (1973) and Lapage *et al.* (1970, 1973).

Alternatively, codes may be given to the set of results. These can be semi-automated or determined manually by giving a single code number to each of several sets of tests, thus obtaining a series of code numbers (e.g. 5−8 numbers). This type of approach is discussed by D'Amato *et al.* (1981) and is used by several systems including the API 20 series and the Minitek system. Development of codes for plant pathogens would rely on the use of a wide range of strains for each taxon. Initial use of the API 50 series would allow rapid selection of key tests for use with the desired group of bacteria. Appropriate kits could then be developed either from basic materials using microtitre plates or using commercially obtainable impregnated discs. Once the cultural conditions of the test have been established, tests of a wide range of pathogens would allow development of manual or computerized databases from which accurate identification of bacteria with discrete nutritional profiles could be made. Although such methods are often very useful for identification to specific level they are often inappropriate for accurate identification at infraspecific levels especially, for example, within *Pseudomonas syringae* and *Xanthomonas campestris* pathovars.

The use of numerical analytical techniques for plant pathogenic bacteria is reviewed by Goor *et al.* (1990). Details of some existing programmes for use in taxonomy and identification may be obtained from Sackin (1987).

Identification of bacteria using whole cell fatty acid profiles

The lipid moieties of the bacterial cell membrane and lipopolysaccharide comprise fatty acids. Fatty acid analysis is increasingly being used as a taxonomic tool to show similarities and differences between groups of bacteria in studies describing new taxa or proposing nomenclatural changes. For many groups there is a good correlation between classification based on fatty acid analysis and on nucleic acid homology. Classification based on fatty acids provides the basis for rapid and accurate identification of many bacteria even at infraspecific level, and fatty acid profiles are now considered to be the most rapid and accurate techniques available to identify bacteria (Moss *et al.*,

1980; Stead, 1988). Fatty acids in bacteria are reviewed by Ratledge & Wilkinson (1988, 1989) and also by Lechevalier (1977).

Most bacterial fatty acids contain 9–20 carbon atoms. The major types are illustrated in Fig. 5.1. Some bacterial fatty acids may belong to more than one group, for example, some branched fatty acids may also have a hydroxy group. To-date over 200 fatty acids have been identified in bacteria and this great diversity provides the basis of a diagnostic profile which is based on both qualitative and quantitative differences.

Fatty acid profiling comprises a number of discrete stages.

1 Cell culture under standard conditions.

2 Harvesting cells.

3 Saponification of lipids.

4 Methylation of fatty acids.

5 Extraction of fatty acid methyl esters (FAMEs).

6 Purification.

7 Separation of FAMEs by gas chromatography (GC).

8 Identification and quantificaton of FAME peaks using known calibration standards.

9 Comparison of the profile with a library of known profiles.

In its simplest form, fatty acid profiling involves manual identification of GC peaks by comparison with known standards and calculation of peak area as a percentage of the total peak

Straight chain saturated

$CH_3-CH_2-CH_2 \text{-----} CH_2-COOH$ — Examples: 16:0

Straight chain unsaturated

$CH_3-CH_2-CH_2-CH=CH-CH_2 \text{---} CH_2-COOH$ — 16:1 *cis* 9

$CH_3-CH_2-CH=CH \text{--} CH_2-CH=CH-CH_2 \text{---} CH_2-COOH$ — 16:2 *trans* 9, 12

Branched

$CH_3 \searrow$
$\quad CH-CH_2-CH_2 \text{------} COOH$ — 15:0 *iso*
$CH_3 \nearrow$

$CH_3-CH_2 \searrow$
$\qquad CH-CH_2-CH_2 \text{------} COOH$ — 15:0 *anteiso*
$CH_3 \nearrow$

Cyclopropanes

$CH_3-CH_2-CH-CH-CH_2 \text{---} COOH$
$\qquad\quad \backslash \; / $
$\qquad\quad CH_2$ — 17:0 *cyclo*

Hydroxy

$CH_3-CH_2-CH_2 \text{-----} CHOH-COOH$ — 16:0 2OH

$CH_3-CH_2-CH_2 \text{-----} CHOH-CH_2-COOH$ — 16:0 3OH

Fig. 5.1 Major types of fatty acids found in plant pathogenic bacteria.

area. Identification at genus level can often be predicted from the types of fatty acid present. In this respect the hydroxy acids are of great importance in the Gram negative bacteria. Many species can be differentiated on the basis of ratios of selected fatty acids. Since most fatty acids in the profile are of some diagnostic value, comparison of a whole profile with a library entry representing a wide range of strains of a particular taxon offers a potentially more accurate diagnostic system, but computerized cataloguing and pattern recognition systems are required. Taken to its extreme there is the possibility of a fully automated system in which a test profile is compared with a library of all known taxa contained on a computer database (Microbial ID Inc.).

There are as yet relatively few published papers on fatty acid profiling of plant pathogenic bacteria. However, the following references give an introduction: Oyaizu & Komagata (1983); Sasser & Miller (1984); van der Zwet & Sasser (1985); De Boer & Sasser (1986); Roy (1988) and Stead (1988, 1989).

Cultural techniques
It is essential to consider to what use profiles may be put before deciding on a culture medium. If the study is a one-off examination of a small group of bacteria then a medium which gives good growth should be chosen. If the aim is to develop a large library of profiles of a wide range of bacteria against which unknowns can be compared then the medium should be chemically defined, give good reproducibility from batch to batch and support good, rapid growth of a wide range of species. This medium should contain very few fatty acids which may interfere with the bacterial profile (in case small quantities of medium are accidently carried over). Trypticase soy agar (TSA), King's medium B and 1% glucose nutrient agar may all be used for plant pathogenic bacteria. In my laboratory TSA is used because it suits a wide range of bacteria, it is easy and cheap to prepare, there is little or no carry over of fatty acids from the medium and one of the commercially available libraries (TSBA Aerobic Library from Microbial ID Inc.) is based on growth on this medium.

Profiles will vary from medium to medium, mainly through quantitative differences for certain fatty acids or groups of fatty acids. Thus, for some closely related taxa, the effect of the medium composition may be greater than the differences between the profiles of the species produced from the same medium. The same applies to incubation temperature and time. Generally for Gram negative bacteria increase in temperature increases the relative amounts of saturated FAMEs and decreases unsaturated FAMEs. Increase in incubation periods for fluorescent

Pseudomonas spp. increases the relative amounts of cyclo-propane acids. Since fatty acid profiling is often used for rapid, accurate diagnosis, a shorter incubation period is preferred but profiles often become more stable during the late log growth phase. For most plant pathogenic and other bacteria a temperature of 28°C for 24 h is adequate. However, for *Xanthomonas* spp. 48-h TSA cultures are more useful than 24-h cultures. Not only is it easier to harvest sufficient cells but profiles seem to be more stable and give more reliable differentiation between pathovars of *X. campestris*. I also use 48-h cultures for some other slow-growing bacteria, e.g. *Clavibacter michiganensis* subsp. *sepedonicus* and *Xylophilus ampelinus*.

The quantity of cells harvested and methods of streaking the plate are two other important factors which need to be standardized. Since the physiological age of cells can influence the profile it is best to streak the plate in four quadrant streaks so that isolated colonies are found in quadrant four. Avoid the wash-out streaks in quadrant one and the selection of isolated colonies in other quadrants. Select the confluent growth from quadrants two and three (late log phase after 24-h growth) and combine this growth from two or more plates rather than scrape up the cells from one plate. The best extractions using the method described will be given by 30−60 mg fresh weight of cells. Below this range some of the minor acids (less than 1% of the total peak area) which are often of great diagnostic value are likely to be diluted out. Above 60 mg there is an increased chance that gels may develop during processing and that extraction efficiency for some acids may be reduced.

However, if you decide to generate your own library for a particular group of bacteria it is good practice to vary the culture media including sterile medium controls, to investigate the effect of incubation temperature and time, harvesting techniques and cell fresh weight on the profiles. Only by understanding the effects of these parameters will you have confidence in your ability to differentiate between pathogens. Comprehensive details of the methods, based on those of Miller and Berger (1985), for culturing the bacteria through to separation and identification of FAMEs are given in Appendix 3.

Identification of bacteria by gel electrophoretic profiles of whole cell proteins

Since the genetic information held in the DNA is translated into proteins it is highly likely that a profile of these proteins will be specific to a particular taxon and perhaps even a particular strain. It is estimated that approximately 2000 genes may be expressed as different proteins in a bacterial cell

(Jackman, 1987). If a profile containing 2000 proteins could be obtained and assuming quantitative analysis was also possible, such profiles should allow very accurate identification. In this section the methods available for obtaining protein profiles are discussed; it also aims to highlight some of the practical problems, to describe in detail a standard method, and to review the ways in which the profile obtained can be used to identify a bacterium.

Although protein profiles have potential in the rapid identification of bacterial cultures, they are not widely used for routine identification of plant pathogenic bacteria, although they have been used in taxonomic studies of the genus *Xanthomonas*. The major reasons for this are the complexity of the profiles and the problems with inter- and intrageneric standardization. Although such analysis can be computerized it is time-consuming compared with other methods of identification which take 1–2 days. At present the technique does not lend itself to identification of large numbers of cultures. Software has been developed which considerably reduces the time taken to standardize profiles but it is not yet commercially available.

There are relatively few publications on the use of these techniques for identification of plant pathogenic bacteria (Davis *et al.*, 1984; Kersters *et al.*, 1989).

Electrophoretic techniques

Cellular proteins can be separated by several electrophoretic techniques whereby the protein molecules in solution migrate in response to an electric current. The rate at which they migrate depends upon several factors:
1 the strength of the electrical field;
2 the net charge, size and shape of the proteins;
3 the ionic strength, viscosity and temperature of the supporting medium.

For a general review of electrophoretic techniques for proteins see Hames & Rickwood (1981).

Bacterial proteins are commonly separated by two techniques: (i) sodium dodecyl sulphate-polyacrylamide gel electrophoresis (SDS-PAGE); and (ii) agarose gel isoelectric focusing (IEF). Both are usually performed in one dimension. Two-dimensional electrophoresis could give a more accurate analysis but it tends to be impractical, largely through difficulties of standardization.

In SDS-PAGE, SDS binds to the proteins in such a way that they have uniform hydrodynamic and charge characteristics so that all the proteins in a sample migrate to the anode at a rate inversely proportional to their molecular weight. Small proteins travel furthest in the gel. The gel itself also has a sieving effect. In contrast, proteins are separated in IEF on the basis of their net charge. Thus, IEF requires a gel with a much larger pore size. Pore size can be adjusted for both polyacrylamide and agarose gels and they are prepared either as slabs or as tubes. For bacterial proteins the most common method is SDS-PAGE using slabs of about 14 cm × 16 cm × 1.5 mm.

The use of SDS-PAGE in microbial systematics has been reviewed by Jackman (1987) who proposes three limiting factors as the reasons why relatively few people use protein profiling for identification: (i) lack of an accepted single technique; (ii) problems with reproducibility in electrophoretic techniques; and (iii) lack of objective methods for comparison of profiles. He proposed a method based on that of Laemmli (1970) as the standard protein profiling technique for systematic and identification studies (Jackman, 1985; Kersters, 1990; Appendix 4).

Cultural conditions   The range of species in the study group should be defined and a medium selected which gives good growth of all representatives. A standard temperature and incubation period should be selected.

Extraction of proteins   There are three possible approaches and the final choice will vary with the type of bacterium under study as bacteria vary greatly in their susceptibility to cell disruption methods.
1 Physical methods, e.g. heat or ultrasound. These methods are not recommended when dealing with potential human pathogens.
2 Chemical methods, e.g. treatment with SDS, usually combined with heat. This is perhaps the most common method for Gram negative bacteria and is described in Appendix 4.
3 Biological methods, e.g. treatment with lysozyme, often as a pretreatment before disruption with SDS.

(SDS-PAGE)   Full details of the preparation and running of the gels are given in Appendix 4.

Analysis of patterns   Analysis may be made by eye but the complexity of profiles most often necessitates analysis of densitometer traces using a computer programme. Computerized analyses are discussed by Kersters & De Ley (1975); Feltham & Sneath (1979); Jackman *et al.* (1983) and Jackman (1985, 1987). The basic strategy for analysis includes the following:
1 conversion of band pattern into a densitometer trace;
2 correction of the trace for intergel differences;
3 calculation of similarity between traces;
4 formation of a similarity matrix which will cluster strains to show the interrelationships of the traces analysed;

5 development of a library of profiles based on the above matrix;
6 identification of unknowns by comparison with the library.
For some closely related taxa such as *P. syringae* pathovars,
the gel-to-gel variation may be as great as the real variation in
profiles. They are difficult to differentiate unless conditions are
standardized; a reference strain and molecular weight markers
should be included on each gel.

Summary

The key features of a successful diagnostic test are accuracy,
minimal labour input per test, speed of test and cost. Of course,
the perfect test should be rapid, accurate and cheap.

When considering the speed of result it is tempting to avoid
tests which rely on isolation and purification of the pathogen
from the natural host and subsequent identification of the culture. However, it must be remembered that isolation onto a
good selective medium is still often the most sensitive method
for detecting a pathogen in diseased tissues or in latent infections. In many instances it may in any case be necessary to
confirm the diagnosis by inoculating a culture to an indicator
host plant.

Traditional nutritional and physiological tests require a pure
culture. They are usually accurate at the species level but are
often very inaccurate at the infraspecific level, especially for
pathovar determination when host plant tests are usually
required for confirmation. They do not use expensive reagents,
chemicals or equipment but do require a large input of labour
time and the result is often not available for several weeks.
Nutritional profiles using kits have the same limitations in
terms of accuracy and still need a pure culture. Also, most are
not designed for plant pathogenic bacteria.

Serological tests can be used for accurate identification of
cultures but their greatest potential is in rapid detection of the
pathogen either in diseased tissue, subclinical infections or
seed samples. The most important feature of any serological
test is the quality (specificity) of the antiserum employed. The
tests themselves merely amplify or increase the visibility of the
antigen−antibody reaction. The IF test is probably the best
method for plant pathogenic bacteria. It is sensitive ($10^3 - 10^4$
cells ml^{-1}) and the serological reaction can be seen on individual
cells, thus avoiding many false positive results. The major
disadvantage is the time taken to examine each microscope
slide and the limited numbers of tests that can be examined in
a given time. ELISA is much more suitable for testing large
numbers of samples but the present tests tend to be less sensi-

tive (10^5-10^6 cells ml^{-1}) than IF tests and false positive results are more common.

Fatty acid profiles must also be done on a pure culture but this is probably the best single technique for identifying bacteria and, coupled with good selection of colonies from isolation plates, it can be a very accurate and rapid method. It is almost always accurate at the species level and can also identify many bacteria at the infraspecific level, especially *Xanthomonas campestris* pathovars. It requires costly equipment and software but this is offset against low staff costs. Labour input per profile is about 6 min. With an automated system 200–250 tests per week can be done by one person which makes it cost effective for large-scale surveys and it will become more useful as the computer database libraries of profiles increase in size.

The technique of protein profiling in SDS-PAGE gels also has potential for identification of cultures but the present problems of standardization of electrophoretic separations and analysis of complex band patterns make the technique more suitable for identification of small numbers of samples.

References

Alvarez A.M., Benedict A.A. & Mizumoto C.Y. (1985) Identification of xanthomonads and grouping of *Xanthomonas campestris* pv. *campestris* with monoclonal antibodies. *Phytopathology* **75**, 722–8.

Alvarez A.M., Benedict A.A., Mizumoto C.Y. & Civerolo E.L. (1987) Mexican lime bacteriosis examined with monoclonal antibodies. In Civerolo E.L., Collmer A., Davis R.E. & Gillaspie A.G. (eds) *Plant Pathogenic Bacteria*, pp. 845–52. Martinus Nijhoff, Dordrecht.

Aquino T.I. & Dowell L. (1975) A comparison of Mini-Tek (BBL) and API (Analytab) for identification of Enterobacteriaceae. *Abstracts of the Annual Meeting of the American Society for Microbiology*, p. 28. American Society for Microbiology, Washington.

Austin B. & Priest F. (1986) *Modern Bacterial Taxonomy*. van Nostrand Reinhold, Wokingham.

Barbara D.J. & Clark M.F. (1982) A simple indirect ELISA using F(ab')$_2$ fragments of immunoglobulin. *Journal of General Virology* **58**, 315–22.

Barnett O.W. (1986) Application of new test procedures to surveys: merging the new with the old. In Jones R.A.C. & Torrance L. (eds) *Developments and Applications in Virus Testing*, pp. 247–67. Association of Applied Biologists, Wellesbourne.

Bascomb S., Lapage S.P., Curtis M.A. & Willcox W.R. (1973) Identification of bacteria by computer: identification of reference strains. *Journal of General Microbiology* **77**, 291–315.

Benedict A.A., Alvarez A.M., Berestecky J. *et al.* (1989) Pathovar specific monoclonal antibodies for *Xanthomonas campestris* pv. *oryzae* and for *Xanthomonas campestris* pv. *oryzicola*. *Phytopathology* **79**, 322–8.

Benedict A.A., Alvarez A.M., Mizumoto C.Y. & Civerolo E.L. (1985) Delineation of *Xanthomonas campestris* pv. *citri* strains with monoclonal antibodies. *Phytopathology* **75**, 1352.

Bonner R.L., Alvarez A.M., Berestecky J. & Benedict A.A. (1987) Monoclonal antibodies used to characterise *Xanthomonas campestris* pv. *dieffenbachiae.* *Phytopathology* **77**, 1725.

Brennan K.A., Ellner P.D. & Kiehn T.E. (1974) Evaluation of the Minitek system for identifying Enterobacteriaceae. *Abstracts of the Annual Meeting of the American Society for Microbiology*, p. 130. American Society for Microbiology, Washington.

Burdash N.M., Bannister E.R., Manos J.P. & West M.E. (1980) A comparison of four commercial systems for the identification of nonfermentative Gram-negative bacilli. *American Journal of Clinical Pathology* **73**, 564–69.

Chirkov S.N., Olovnikov A.M., Surgachyova H.A. & Atabekov J.G. (1984) Immunodiagnosis of plant viruses by a virobacterial agglutination test. *Annals of Applied Biology* **104**, 477–83.

D'Amato R.F., Holmes B. & Bottone E.J. (1981) The systems approach to diagnostic microbiology. *CRC Critical Reviews in Microbiology* **9**, 1–44.

Davis M.J., Gillespie A.G., Vidaver A.K. & Harris R.W. (1984) *Clavibacter*: a new genus containing some phytopathogenic coryneform bacteria including *Clavibacter xyli* subsp. *xyli* sp. nov., subsp. nov. and *Clavibacter xyli* subsp. *cynodontis* subsp. nov., pathogens that cause ratoon stunting disease of sugarcane and Bermudagrass stunting disease. *International Journal of Systematic Bacteriology* **34**, 107–17.

De Boer S.H. & McCann M. (1989) Determination of population densities of *Corynebacterium sepedonicum* in potato stems during the growing season. *Phytopathology* **79**, 946–51.

De Boer S.H. & McNaughton M.E. (1986) Evaluation of immunofluorescence with monoclonal antibodies for detecting latent bacterial ring rot infections. *American Potato Journal* **63**, 533–43.

De Boer S.H. & McNaughton M.E. (1987) Monoclonal antibodies to the lipopoly-saccharide of *Erwinia carotovora* subsp. *atroseptica* serogroup 1. *Phytopathology* **77**, 828–32.

De Boer S.H. & Sasser M. (1986) Differentiation of *Erwinia carotovora* ssp. *carotovora* and *E. carotovora* spp. *atroseptica* on the basis of cellular fatty acid composition. *Canadian Journal of Microbiology* **32**, 796–800.

De Boer S.H. & Wieczorek A. (1984) Production of monoclonal antibodies to *Corynebacterium sepedonicum. Phytopathology* **74**, 1431–4.

De Boer S.H., Wieczorek A. & Kummer A. (1988) An ELISA test for bacterial ring rot of potato with a new monoclonal antibody. *Plant Disease* **72**, 874–8.

Edwards M.L. & Cooper J.I. (1985) Plant virus detection using a new form of indirect ELISA. *Journal of Virological Methods* **11**, 309–19.

Fahey P.C. & Persley G.J. (1983) *Plant Bacterial Diseases: A Diagnostic Guide.* Academic Press, London.

Feltham R.K.A. & Sneath P.H.A. (1979) Quantitative comparison of electro-phoretic traces of bacterial proteins. *Computers and Biomedical Research* **12**, 247–63.

Goor M., Kersters K., Mergaert J. *et al.* (1990) Numerical analysis of phenotypic features. In Klement Z., Rudolph K. & Sands D.C. (eds) *Methods in Phyto-bacteriology*, pp. 145–52. Akademiai Kiado, Budapest.

Hames B.D. & Rickwood D.Y. (1981) *Gel Electrophoresis of Proteins: A Practical Approach.* IRL Press, Oxford.

Hansen S.L., Hardesty D.R. & Myers B.M. (1974) Evaluation of the BBL Minitek System for the identification of Enterobacteriaceae. *Applied Micro-biology* **28**, 798–801.

Hanson C.W., Cassorla R. & Martin W.J. (1979) API and Minitek systems in identification of clinical isolates of anaerobic Gram-negative bacilli and *Clostridium* species. *Journal of Clinical Microbiology* **10**, 14–18.

Hanson C.W. Marso E. & Martin W.J. (1978) Comparison of the Minitek test system with a conventional screening procedure for identification of Entero-bacteriaceae. *Health Laboratory Science* **15**, 3–8.

Hayward A.C., El-Nashaar H.M., De Lindo L. & Nydegger V. (1989) The use of microtiter plates in the phenotypic characterisation of phytopathogenic pseudomonads. In Klement Z. (ed.) *Proceedings of the 7th International Conference on Plant Pathogenic Bacteria*, part A, pp. 593–8. Akademiai Kiado, Budapest.

Holmes B. & Humphry P.S. (1988) Identification of Enterobacteriaceae with the Minitek system. *Journal of Applied Bacteriology* **64**, 151–61.

Jackman P.J.H. (1985) Bacterial taxonomy based on electrophoretic whole cell protein patterns. In Goodfellow M. & Minnikin D. (eds) *Chemical Methods in Bacterial Systematics*, pp. 115–29. Academic Press, London.

Jackman P.J.H. (1987) Microbial systematics based on electrophoretic whole-cell protein patterns. In Colwell R. & Grigorova R. (eds) *Methods in Microbiology*. Vol. 19, pp. 209–25. Academic Press, London.

Jackman P.J.H., Feltham R.K.A. & Sneath P.H.A. (1983) A programme in BASIC for numerical taxonomy of microorganisms based on electrophoretic protein patterns. *Microbios Letters* **23**, 87–98.

Kersters K. (1990) Polyacrylamide gel electrophoresis of bacterial proteins. In Klement Z., Rudolph K. & Sands D.C. (eds) *Methods in Phytobacteriology*, pp. 191–8. Akademiai Kiado, Budapest.

Kersters K. & De Ley J. (1975) Identification and grouping of bacteria by numerical analysis of their electrophoretic protein patterns. *Journal of General Microbiology* **87**, 333–42.

Kersters K., Pot B., Hoste B., Gillis M. & De Ley J. (1989) Protein electrophoresis and DNA: DNA hybridisations of xanthomonads from grasses and cereals. *EPPO Bulletin* **19**, 51–5.

Klement Z., Rudolph K. & Sands D.C. (1990) *Methods in Phytobacteriology*. Akademiai Kiado, Budapest.

Laemmli U.K. (1970) Cleavage of structural proteins during the assembly of the head of bacteriophage T4. *Nature* **227**, 680–5.

Lapage S.P., Bascomb S., Willcox W.R. & Curtis M.A. (1970) Computer identification of bacteria. In Baillie A. & Gilbert R.J. (eds) *Automation, Mechanisation and Data Handling in Microbiology*, pp. 1–22. Society for Applied Bacteriology Technical Series No. 4, Academic Press, London.

Lapage S.P., Bascomb S., Willcox W.R. & Curtis M.A. (1973) Identification of bacteria by computer: general aspects and perspectives. *Journal of General Microbiology* **77**, 273–90.

Lechevalier M.P. (1977) Lipids in bacterial taxonomy — a taxonomist's view. *CRC Critical Reviews in Microbiology* **5**, 109–210.

Lelliott R.A. & Stead D.E. (1987) *Methods for the Diagnosis of Bacterial Diseases of Plants*. Blackwell Scientific Publications, Oxford.

Macario A.J.L. & De Macario E.C. (1988) Monoclonal antibodies against bacteria. *Biotechnological Advances* **6**, 135–50.

Maestroeni P., Carbone M., Fera M.T., Teri G. & Burdash N.M. (1984) Comparison of six commercial systems for the identification of Enterobacteriaceae. *Public Health Laboratory* **42**, 50–63.

Magee W.E., Beck C.F. & Ristow S.S. (1986) Monoclonal antibodies specific for *Corynebacterium sepedonicum*, the causative agent of potato ring rot. *Hybridoma* **5**, 231–5.

Mergaert J., Verdonck L., Kersters K., Swings J., Boefgras J.M. & De Ley J. (1984) Numerical taxonomy of *Erwinia* species using API systems. *Journal of General Microbiology* **130**, 1893–910.

Miller L. & Berger T. (1985) Bacteria identification by gas chromatography of whole cell fatty acids. *Hewlett Packard Gas Chromatography Application*, Note 228–41, 8 pp. Hewlett Packard, USA.

Moss C.W., Dees S.B. & Guerrans G.O. (1980) Gas chromatography of bacterial fatty acids with a fused silica column. *Journal of Clinical Microbiology* **12**, 127–30.

Nord C.E., Lindberg A.A. & Dahlback A. (1974) Evaluation of five test kits: API, Auxtotab, Enterotube, Pathotec and R/B − for identification of Enterobacteriaceae. *Medical Microbiology and Immunology* **159**, 211−20.

Oyaizu H. & Komagata K. (1983) Grouping of *Pseudomonas* species on the basis of cellular fatty acid composition and the quinone system with special reference to 3-hydroxy fatty acids. *Journal of General and Applied Microbiology* **29**, 17−40.

Ratledge C. & Wilkinson S.G. (1988) *Microbial Lipids*, Vol. 1. Academic Press, London.

Ratledge C. & Wilkinson S.G. (1989) *Microbial Lipids*, Vol. 2. Academic Press, London.

Roy M.A. (1988) Use of fatty acids for the identification of phytopathogenic bacteria. *Plant Disease* **72**, 460.

Ruissen M.A., Helderman C.A., Schipper J. & van Vuurde J.W.L. (1986) Selective isolation and concentration of phytopathogenic bacteria on immunoaffinity columns. In Civerolo E.L., Collmer A., Davis R.E. & Gillaspie A.G. (eds) *Plant Pathogenic Bacteria*, pp. 882. Martinus Nijhoff, Dordrecht.

Sackin M.J. (1987) Computer programs for classification and identification. In Colwell R. & Grigorova R. (eds) *Methods in Microbiology*, Vol. 19, pp. 459−94. Academic Press, London.

Sasser M. & Miller L. (1984) Identification of pseudomonads by fatty acid profiling. In *Proceedings of the 2nd Working Group on Pseudomonas syringae pathovars*, pp. 45−7. The Hellenic Phytopathological Society, Sounion.

Sasser M. & Smith D.H. (1987) Parallels between ribosomal RNA and DNA homologies and fatty acid composition in *Pseudomonas* (abstr.). *Abstracts of the 87th Annual Meeting of the American Society for Microbiology*, p. 241. American Society for Microbiology, Washington.

Schaad N.W. (1988) *Laboratory Guide for Identification of Plant Pathogenic Bacteria*, 2nd edn. American Phytopathological Society, St. Paul, Minnesota.

Schaad N.W., Sule S., van Vuurde J.W.L. *et al.* (1990) Serology. In Klement Z., Rudolph K. & Sands D.C. (eds) *Methods in Phytobacteriology*, pp. 153−90. Akademiai Kiado, Budapest.

Shayegani M., Hubbard M.E., Hiscott T. & McGlynn D. (1975) Evaluation of the R/B and Minitek systems for identification of Enterobacteriacae. *Journal of Clinical Microbiology* **1**, 504−8.

Slifkin M., Gil G.M. & Engwall C. (1986) Rapid identification of Group K and other corynebacteria with the Minitek system. *Journal of Clinical Microbiology* **24**, 177−80.

Smith E.G., Pritchard J.K. & McCarthy L.R. (1981) Four-hour presumptive identification of Enterobacteriaceae from blood cultures. *American Journal of Clinical Pathology* **75**, 81−91.

Stead D.E. (1987) Immunofluorescence techniques in plant pathology. In Grange J.M., Fox A. & Morgan N.L. (eds) Vol. 24, pp. 129−36. Society for Applied Bacteriology Technical Series, Blackwell Scientific Publications, Oxford.

Stead D.E. (1988) Identification of bacteria by fatty acid profiling. *Acta Horticulturae* **225**, 39−46.

Stead D.E. (1989) Grouping of *Xanthomonas campestris* pathovars of cereals and grasses by fatty acid profiling. *EPPO Bulletin* **19**, 57−68.

Stead D.E., Chauveau J.F., Janse J.D., Ruissen M.A., van Vaerenbergh, J. and van Vuurde J.W.L. (1987) Immunoisolation techniques for the detection and isolation of plant pathogenic bacteria. In Grange J.M., Fox A. & Morgan N.L. (eds) *Immunological Techniques in Microbiology*, Vol. 24, p. 189−93. Society for Applied Bacteriology Technical Series, Blackwell Scientific Publications, Oxford.

van der Zwet T. & Sasser M. (1985) Characterisation of *Erwinia amylovora* through fatty acid profiling. *Phytopathology* **75**, 1281.

van Regenmortel M.H.V. (1982) *Serology and Immunochemistry of Plant Viruses.* Academic Press, London.

van Vuurde J.W.L. (1987) New approach in detecting phytopathogenic bacteria by combined immunoisolation and immunoidentification assays. *EPPO Bulletin* **17**, 139–48.

van Vuurde J.W.L. & van Henten C. (1983) Immunosorbent immunofluorescence microscopy (ISIF) and immunosorbent dilution plating (ISDP): new methods for the detection of plant pathogenic bacteria. *Seed Science and Technology* **11**, 525–33.

van Vuurde J.W.L., Ruissen M.A. & Vruggink H. (1986) Principles and prospects of new serological techniques including immunosorbent immunofluorescence, immunoffinity isolation and immunosorbent enrichment for sensitive detection of phytopathogenic bacteria. In Civerolo E.L., Collmer A., Davis R.E. & Gillaspie A.G. (eds) *Plant Pathogenic Bacteria*, pp. 835–42. Martinus Nijhoff, Dordrecht.

Villagarcia N. (1985) The scope and limitations of four methods for the rapid identification of Enterobacteriaceae in foods. *Journal of Applied Bacteriology* **58**, 123–9.

Weir D.M. (ed.) (1978) *Handbook of Experimental Immunology.* Blackwell Scientific Publications, Oxford.

Wellstood-Nuesse S. (1979) Comparison of the Minitek system with conventional methods for identification of nonfermentative and oxidase-positive fermentation Gram-negative bacilli. *Journal of Clinical Microbiology* **9**, 511–16.

Yuen G.Y., Alvarez A.M., Benedict A.A. & Trotter K.J. (1987) Use of monoclonal antibodies to monitor the dissemination of *Xanthomonas campestris* pv. *campestris. Phytopathology* **77**, 366–70.

Appendix 1: Immunofluorescence techniques

The indirect IF protocol for identification of bacteria on slides is as follows.

Buffers
- 0.05 M phosphate-Tween buffer, pH 7.0

Na_2HPO_4	4.26 g
KH_2PO_4	2.72 g
Distilled water	1.0 litre
Tween 20	0.5 ml

- 0.01 M phosphate buffer, pH 7.2

$Na_2HPO_4.12H_2O$	2.7 g
$NaH_2PO_4.2H_2O$	0.4 g
Distilled water	1.0 litre

- 0.1 M phosphate buffered glycerine, pH 7.6

$Na_2HPO_4.12H_2O$	3.2 g
NaH_2PO_4	0.15 g
Glycerine	50 ml
Distilled water	100 ml

Method
Figure 5.2 shows suggested layouts for determination of titres and for performing IF tests on plant tissue extracts on slides. Always set up a positive control slide with homologous antigen for comparative purposes. For each window do the following:
- Add bacterial suspension or plant tissue extract at appropriate

(a) Dilutions of antiserum

FITC   10   20   40   80   160

Standard
dilution
of known
bacterium

○ ○ ○ ○ ○ ○
○ ○ ○ ○ ○ ○

320   640   1280   2560   5120   10 240

(b) Dilutions of known bacterium

Neat   10   100   1000

Standard
dilution
of
antibody

○ ○ ○ ○ ○ ○
○ ○ ○ ○ ○ ○

Duplicate

FITC

(c) Dilutions of macerated plant tissue

Neat   10   100   1000

Standard
dilution
of
antibody

○ ○ ○ ○ ○ ○
○ ○ ○ ○ ○ ○

Duplicate

FITC

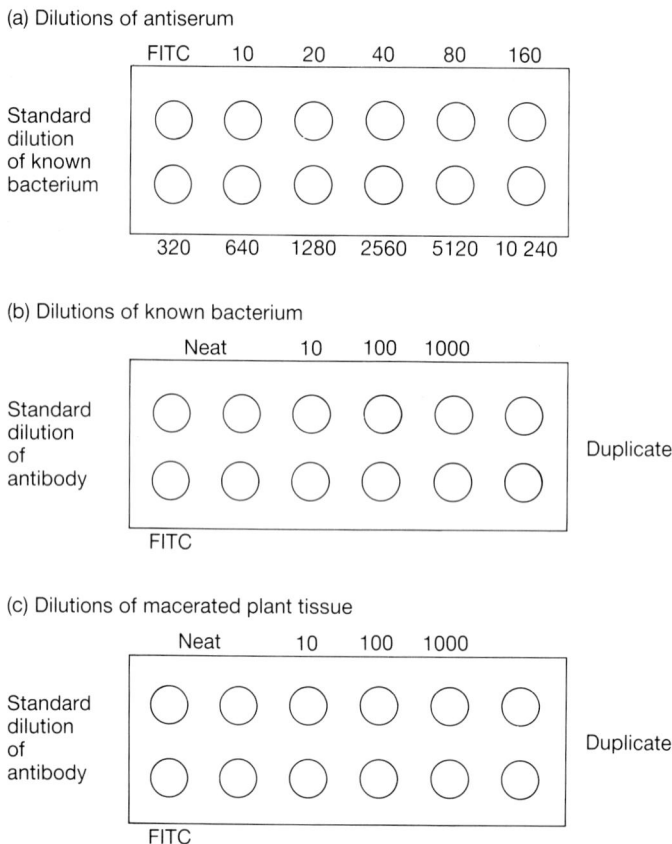

Fig. 5.2 Layouts for IF tests on plant tissue extracts on slides.

dilution in 0.05 M phosphate-Tween buffer, pH 7.0, to the window (25 µl for a 6-mm-diameter window) and allow to air dry at 37°C or in a laminar airflow cabinet.
• Fix by gently heating in a flame or dipping in 3% formalin in phosphate buffer for 10 min. Wash in phosphate buffer for 5 min.
• Add sufficient antiserum at the appropriate dilution (found by experiment) to cover the entire window (about 25 µl). Cover one window with buffer instead of antiserum (FITC conjugate control). Incubate for 30 min at room temperature in a humid chamber.
• Rinse off the antiserum and wash three times for 3 min each in phosphate buffer in a Coplin jar. Keep positive control slides separate to avoid cross-contamination.
• Remove excess moisture with blotting paper, avoiding removal of preparation and cross-contamination between windows.
• Add sufficient FITC conjugate at the recommended working dilution to cover each window. Incubate again in a humid chamber; rinse and wash as above in phosphate buffer.
• Remove excess moisture.

- Add $5-10\,\mu l$ of phosphate buffered glycerine. Cover with a large coverslip and observe. Scan two window diameters at right angles and approximately one radian along the edge of the window. Observe for bright green peripheral fluorescence of cells. Check the control windows first. If fluorescent cells are found in the FITC conjugate control window, repeat the whole procedure and, if still found, discard the slides and use a different antiserum or conjugate. If you wish to make a count, take the mean of at least 20 randomly selected fields of view from each window.

Appendix 2: Detection of plant pathogenic bacteria by ELISA

Buffers

- 0.01 M phosphate buffered saline (PBS), pH 7.4

NaCl	8 g
KH_2PO_4	0.2 g
$Na_2HPO_4.12H_2O$	2.9 g
KCl	0.2 g
Distilled water	1.0 litre

- PBS-Tween

  PBS containing 0.05% Tween 20
- Carbonate coating buffer, pH 9.6

Na_2CO_3	1.59 g
$NaHCO_3$	2.93 g
Distilled water	1.0 litre

- Sample buffer

PVP mol. wt 44 000	2.0 g
Skimmed milk blocking agent	2.0 ml
Tween 20	50 μl
PBS	100 ml

- Skimmed milk blocking agent

Skimmed milk powder	10.0 g
NaN_3	0.2 g
Distilled water	100 ml

- Diethanolamine substrate buffer, pH 9.8 (for use with alkaline phosphatase)

Diethanolamine	97.0 ml
Distilled water	800 ml

Adjust pH with HCl. Make up to 1 litre with distilled water.

Methods
PTA ELISA

- Coat wells by adding 50 μl poly-L-lysine (1 mg 100 ml^{-1} PBS) to each. Incubate plates at room temperature for 30 min.
- Add 50 μl of test suspension. If a pure culture of bacteria adjust to an optical density of 0.1 at 600 nm. Centrifuge the plates at approximately 5000 g for 10 min. Note that a special plate carrier rotor is required.
- Shake out vigorously and add 50 μl of 0.1% glutaraldehyde solution to each well. Incubate at room temperature for 15 min.
- Shake out vigorously and wash twice with PBS.
- Add 150 μl of 0.1% gelatin solution in PBS to each well and incubate at room temperature for 1 h.

• Plates can be stored overnight at 4°C after adding 300 μl of PBS-Tween.
• Wash plates three times for 3 min each with PBS-Tween.
• Add 100 μl of antiserum (preferably immunoglobulin, Ig fraction) suitably diluted in PBS-Tween.
• Incubate for 3 h at 37°C.
• Wash as before.
• Add 100 μl of anti-Ig−enzyme conjugate at the recommended dilution (appropriate for use with the type of antiserum used).
• Incubate at 37°C for 1 h.
• Wash as before.
• Add 100 μl of substrate solution to each well.
• Incubate under appropriate conditions. For alkaline phosphatase incubate at room temperature for 1 h over a bed of crushed ice.
• Read optical densities at A_{405} using a suitable colorimeter.

ATA ELISA Suitable for soluble antigen.
• Coat plates with 100 μl per well of a suitable dilution of trapping antibody in carbonate coating buffer (e.g. 1:100 dilution). Incubate overnight at 4°C or for 2 h at 37°C.
• Shake out vigorously and wash in three changes of PBS-Tween.
• Prepare sample in sample buffer. For bacterial suspensions use approximately 10^6 cells ml^{-1}. For plant tissue samples triturate approximately 0.5 g in 1 ml buffer. Add to plates at 100 μl per well. Incubate overnight at 4°C or for 1 h at 37°C.
• Wash as before.
• Block by adding 200 μl per well of skimmed milk blocking agent. Incubate for 30 min at 37°C.
• Wash as before.
• Add 100 μl per well of an appropriate dilution of a second antibody. Performance may be improved by diluting the second antibody in skimmed milk blocking agent (1:1000 dilution). Incubate for 1 h at 37°C.
• Wash as before.
• Add 100 μl per well of diluted anti-Ig−enzyme conjugate. Again, performance may be improved by diluting with blocking agent (1:1000 dilution). Incubate for 1 h at 37°C.
• Wash as before.
• Add 100 per well of recommended substrate. Incubate for 1 h at 37°C.
• Read optical densities for each well using suitable colorimeter.

Appendix 3: Identification of bacteria by fatty acid profiling

Recommended protocol for culture and harvest
• 'Train' cultures on trypticase soy agar (TSA) for 12−18 h at 28°C if time allows.
• Streak each strain onto two fresh plates of TSA in a four-quadrant dilution streak covering as much of the plate as possible. For slow-growing bacteria do not flame the loop between streaks; for rapidly growing bacteria flame the loop at least between quadrants one and two. Incubate aerobically at 28°C for 24 ± 2 h.

- Harvest cells by scraping off 50 mg fresh weight with a small aluminium spatula (easily determined by taring the spatula on a sensitive top-pan balance before harvesting) from quadrants two and three. Do not pick up any agar medium. Transfer the cells to the base of a screw-capped test tube (13 cm). Caps should be lined with a Teflon seal. Screw tightly.

Extraction of fatty acids

The cell lipids in the membranes and lipopolysaccaride must first be saponified to release the free fatty acids. Since free fatty acids are difficult to separate by gas chromatography (GC) it is necessary to convert them to their methyl esters. The fatty acid methyl esters (FAMEs) must then be extracted from the cell debris by shaking with a suitable organic solvent. It may be necessary to convert any un-methylated organic compounds to water-soluble forms by washing in sodium hydroxide solution before gas chromatographic analysis.

There are many chemical methods by which FAMEs may be extracted. Some of these are reviewed by Lechevalier (1977). I use the technique described below based on acid methanolysis because it produces profiles of diagnostic value for some very closely related taxa, even down to the pathovar level, and also because the commercial library I use routinely (TBSA Aerobic Library, Microbial ID Inc.) was produced using this method. Again, when comparing profiles for diagnosis it is essential to select a single method and stick to it. However, a range of methods should be investigated for each group of bacteria studied and the one that gives maximum differentiation should be used if appropriate.

There are several potential pitfalls during extraction. The most serious of these can occur during the methylation phase. It is essential in acid methanolysis to use the correct concentration of hydrochloric acid. If the acid is too weak methylation will not occur; if it is too strong certain groups of fatty acids (the cyclopropanes) are likely to be degraded. It is essential to use 6 M HCl (AnalaR grade reagent). General-purpose grade reagent of concentrated HCl (35−40% HCl) is not recommended unless it is accurately adjusted to 6 M (do not assume that conc. HCl is 12 M). If the temperature of methylation exceeds 80°C or if the tubes remain in the water bath for more than 10 min it is likely that the cyclopropane acids will have degraded.

Recommended protocol for saponification, methylation and extraction

- Add 1 ml of sodium hydroxide−methanol solution to each tube (sodium hydroxide 45 g, methanol 150 ml, distilled water 150 ml). Cap tightly. Mix on a vortex mixer for 5−10 s. Place in a boiling water bath for 5 min. Mix again. Replace in the water bath for a further 25 min.
- Methylate the fatty acids by adding 2.0 ml HCl−methanol solution (6.0 M HCl 325 ml, methanol 275 ml). Mix as before. Place in a waterbath at 80 ± 1°C for 10 min. Cool immediately to about 20°C.
- Extract the FAMEs by adding 1.25 ml methyl-tert butyl ether:hexane (1:1, v/v). Rotate tubes vertically for 10 min. Remove all the lower aqueous layer with a Pasteur pipette. There is little chance of cross-contamination so a single pipette may be used.

- Add 3.0 ml of base wash (sodium hydroxide 10.8 g, distilled water 900 ml) and rotate tubes vertically for 5 min to convert any free fatty acids to their sodium salts.
- Carefully pipette off the top two-thirds of the organic layer into a GC vial. Do not transfer any of the aqueous layer or it will damage the GC column. If the emulsion does not separate add one or two drops of saturated sodium chloride solution before pipetting. Seal the GC vials, preferably with Teflon liners.

Separation and identification of FAMEs by gas chromatography

The most important feature of GC separation is the column. Capillary columns are recommended since packed columns tend to give poor recovery of the polar hydroxy acids which are critical in diagnosis of most Gram negative bacteria.

A 25-m fused silica capillary column lined with methyl phenyl silicone is ideal combined with a GC that can increase temperature from 170°C to 270°C at 5°min^{-1}. Hydrogen or helium can be used as the carrier gas and split injection is preferred. Detection is by flame ionization. A suitable integrator is also required.

I use a Hewlett Packard 5890 GC fitted with a Hewlett Packard 25-m fused silica column (5% methyl phenyl silicone). The GC conditions are listed in Table 5.3.

The GC is fitted with an autosampler and integrator and is controlled by a computer programme. The software package includes a FAME peak naming table, a library of profiles and an appropriate pattern recognition algorithm (Microbial ID Inc.). If using such a system you must include the appropriate calibration standard (Microbial ID Inc.) which contains all 9–20 carbon analogues of the basic saturated FAME series plus five hydroxy FAMEs which are used as standards. If you are not using such a fully automated system you must first

Table 5.3 Chromatographic conditions for fatty acid profiling of plant pathogenic bacteria

Model	Hewlett Packard 5890
Mode	Split 1:100
Column liner type	Splitless
Injection volume	1 μl
Detection	FID
Carrier gas: hydrogen 210 kPa	30 ml min^{-1} FID
	55 ml min^{-1} split vent
	5 ml min^{-1} septum purge
Auxilliary gas: nitrogen 140 kPa	30 ml min^{-1} FID
	40 ml min^{-1} trap purge
Air: 280 kPa	400 ml min^{-1} FID
Column head pressure	70 kPa
Injector temperature	250°C
Detector temperature	300°C
Initial column temperature	170°C
Final column temperature	270°C
Rate	5° min^{-1}

FID = flame ionization detection.

develop your own peak naming table or family plot. You should chromatograph a wide range of 9–20 carbon analogues in each of the following series.

FAME type	Example
Straight saturated	9:0, 20:0
Mono unsaturated *cis* $(n-7)$	16:1 *cis* 9, 18:1 *cis* 11
Iso branched	15:0 *iso*, 17:0 *iso*
Anteiso branched	15:0 *anteiso*
Cyclopropanes $(n-7)$	17:0 *cyclo* C9–10, 19:0 *cyclo* C11–12
2-hydroxy	12:0 2OH, 14:0 2OH, 16:0 2OH, 16:1 2OH, 18:1 2OH
3-hydroxy	10:0 3OH 12:0 3OH, 14:0 3OH, 16:0 3OH
Iso-3-hydroxy	11:0 *iso* 3OH, 13:0 *iso* 3OH

Using the following formula calculate the equivalent chain length (*ECL*; relative position at which the molecules elute compared to the two adjacent saturated analogues):

$$ECL_a = (Rt_a - Rt_n)/(Rt_{(n+1)} - Rt_n) + n$$

where Rt_a is the retention time of FAME a; Rt_n is the retention time of C_n:0, the straight chain saturated FAME which precedes FAME a; $Rt_{(n+1)}$ is the retention time of $C_{(n+1)}$:0, the straight chain saturated FAME which follows FAME a.

Plot the integer against the relative length, e.g. if the ECL is 15.4, plot 15 on the x axis and 0.4 on the y axis. Each family should give a straight line plot thus allowing accurate identification of any FAME which falls on the line (Fig. 5.3).

If working with a known group of pathogens, e.g. fluorescent pseudomonads, regular runs of individual FAME standards will allow accurate peak identification. It is essential that chromatographic conditions are standardized and appropriate calibrations performed regularly. Some workers use a GC linked to a mass spectrometer. I tend to use this facility only for those few FAMEs the commercial peak naming table cannot identify or for differentiating FAMEs when two or more co-elute.

Major FAMEs occurring in plant pathogenic bacteria

The genus or group can be determined according to the data in Table 5.4 which includes those hydroxy and other FAMEs that are of most importance in differentiation at this level.

Tables 5.5 and 5.6 show the principal fatty acids and their relative percentages in the profiles of a wide selection of plant pathogenic bacteria; profiles were obtained using the protocol given above.

Agrobacterium: Fatty acid profiles are not influenced by plasmids and therefore correct species differentiation based on virulence genes found on plasmids cannot be obtained with fatty acid profiles. However,

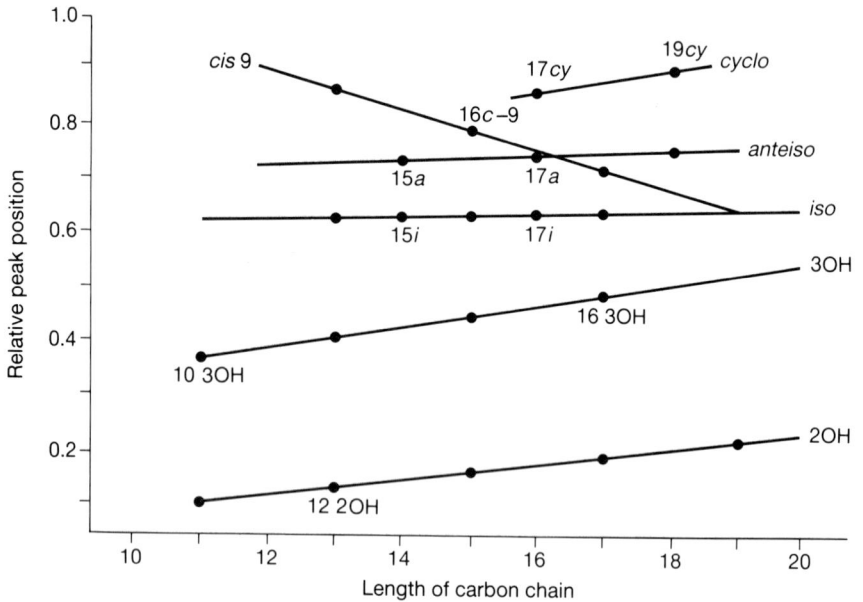

Fig. 5.3 Family plot of equivalent chain lengths for some FAMEs.

differences are found in the biovars. Biovar 2 strains have a unique profile type. There appears to be some overlap between biovar 1 and 3 strains and there are some other strains which do not fit well into the current biovar system; these also are not well differentiated by fatty acid profiling.

Erwinia: This is a genus of convenience created for the plant pathogenic members of the Enterobacteriaceae. All have similar profile types but qualitative and quantitative differences allow accurate differentiation between most species and subspecies. For example, differentiation between *E. chrysanthemi, E. carotovora* subsp. *carotovora* and subsp. *atroseptica* is excellent. Within *E. chrysanthemi* there appear to be some discrete subgroups.

Pseudomonas: There is excellent correlation between classifications based on fatty acid profile and rRNA:DNA hybridization (Sasser & Smith, 1987; Stead, unpublished results). Group 1 is characterized by the presence of 10:0 3OH, 12:0 2OH and 12:0 3OH and contains all fluorescent *Pseudomonas* species plus some non-fluorescent types (*P. aeruginosa, P. amygdali, P. asplenii, P. cichorii, P. corrugata, P. ficuserectae, P. fluorescens, P. fuscovaginae, P. gingeri, P. meliae, P. rubrisubalbicans, P. syringae, P. tolaasii* and *P. viridiflava*). Some *P. syringae* pathovars have unique profiles, e.g. pv. *phaseolicola*, but many overlap.

Table 5.4 Principal fatty acids used to identify plant pathogenic bacteria at the genus level

	8:0 3OH	10:0 3OH	12:0 2OH	12:0 3OH	14:0 3OH	16:1 2OH	16:0 2OH	16:0 3OH	18:1 2OH	11:0 iso 3OH	13:0 iso 3OH	15:1 anteiso A	15:0 iso	15:0 anteiso	17:0 anteiso	10Me 18:0
Agrobacterium					+			+								
Erwinia (Enterobacter)					+											
Pseudomonas rRNA (group 1)		+	(+)	+												
Pseudomonas rRNA (group 2)					+	(+)	(+)	+	+							
Pseudomonas rRNA (group 3)		+														
Xanthomonas				+						+	+		(+)	(+)		
Xylophilus	(+)															
Clavibacter												+	(+)	+	+	
Curtobacterium													+	+	+	
Rhodococcus																+

+ = present in all strains; (+) = present in most strains.

Group 2 is characterized by the presence of 14:0 3OH, 16:0 3OH and 18:1 2OH. Most also have 16:1 2OH and 16:0 2OH. It contains *P. solanacearum, P. andropogonis, P. gladioli, P. cepacia, P. caryophylli, P. plantarii* and *P. glumae*.

Group 3 strains usually have only one hydroxy acid 10:0 3OH although trace amounts of 14:0 3OH have been found in some strains. It contains *P. avenae, P. cattleyae, P. pseudoalcaligenes* subsp. *konjaci* and subsp. *citrulli*, and *P. rubrilineans. P. acidovorans* and *P. testosteroni*, non-plant pathogens included in this rRNA group, are now included in the genus *Comamonas*. Other plant pathogens or plant-associated bacteria currently included in *Pseudomonas* belong to different profile types. These include *P. maltophilia* and *P. cissicola*, which have profiles typical of *Xanthomonas, P. flectens* and *P. paucimobilis*.

Xanthomonas: These bacteria have very distinct profiles containing as many as 50 different fatty acids. Many of these are branched chain acids and almost all strains possess 12:0 3OH, 11:0 *iso* 3OH and 13:0 *iso* 3OH. There is great variation between the profiles of all species and between some pathovars, thus facilitating very accurate determination even at pathovar level, e.g. for *X. campestris* pv. *oryzae*, pv. *oryzicola*, pv. *pelargonii* and pv. *hyacinthii*.

Xylophilus: *X. ampelinus* has a very simple profile characterized by the presence in most strains of 8:0 3OH, an acid not common in any other known plant pathogen.

Table 5.5 Principal fatty acids (%) present in the profiles of some Gram negative plant pathogenic bacteria

	8:0 3OH	10:0 3OH	12:0 2OH	12:0 3OH	14:0 3OH	16:1 2OH	16:0 2OH	16:0 3OH	18:1 2OH	12:0	14:0	16:1 cis 9	16:0	17:0 cyclo	18:1 cis 11	19:0 cyclo 11–12
Agrobacterium biovars					3–10			2–4				2–6	5–10		60–80	1–6
Erwinia amylovora					5–10					4–6	4–6	15–35	30–40	2–15	5–12	
Erwinia carotovora					6–10					4–8	1–3	30–40	20–35		5–25	
Erwinia chrysanthemi					5–10						5–12	20–35	25–35		10–25	
Pseudomonas syringae pv. syringae		1–4	1–5	1–6						2–8		30–40	20–30	0–7	20–25	
Pseudomonas corrugata		2–8	1–7	1–8	1–4					2–7		20–35	22–32	0–12	10–30	
Pseudomonas gladioli					3–6	1–3	1–3	4–6	1–3		3–5	1–25	15–35	2–25	15–35	0–20
Pseudomonas solanacearum					5–10	0–5	0–2		2–6		3–6	25–35	20–35	1–10	10–25	
Pseudomonas avenae	2–5									2–5	0–3	35–45	30–40		15–20	
Xylophilus ampelinus	0–2										2–5	40–50	20–35		15–30	

Table 5.6 Principal fatty acids (%) present in the profiles of some Gram negative and Gram positive plant pathogenic bacteria

	12:0 3OH	11:0 iso 3OH	13:0 iso 3OH	14:0	15:0	16:1 B	16:1 cis 9	16:0	17:B	18:1 cis 9	18:1 cis 11	11:0 iso	14:0 iso	15:1 anteiso A	15:0 iso	15:0 anteiso	16:0 iso	17:1 iso F	17:0 iso	17:0 anteiso	10Me 18:0
Rhodococcus fascians				4–10	5–20			15–30	1–10	5–35											1–13
Clavibacter michiganensis subsp. *michiganensis*								3–8						1–5	0–5	45–55	5–15			27–35	
Clavibacter michiganensis subsp. *sepedonicus*								1–9						1–15	0–2	35–50	5–17		0–2	27–37	
Curtobacterium flaccumfaciens pv. *flaccumfaciens*								0–2			0–10				1–12	45–55	4–13		0–3	20–40	
Xanthomonas maltophilia	1–5	1–6	1–6			1–5	8–15	3–10		1–4	1–5	1–5			30–45	5–20	1–5	2–8	1–10		
Xanthomonas campestris pv. *campestris*	2–6	3–4	2–6			1–4	15–25	2–10				3–6	1–7		20–35	5–20	1–3	1–12	3–10		
Xanthomonas campestris pv. *hyacinthi*		1–3	1–3			1–3	10–15	3–6				2–5			5–25	20–40	5–15	1–5	1–8		
Xanthomonas campestris pv. *oryzae*	2–5		1–3			2–4	15–30	10–30		1–5	1–4	1–4	2–4		1–7	0–2	0–2	5–10	10–20		
Xanthomonas campestris pv. *pelargonii*	2–5	2–3	2–4	1–5		2–3	20–30	2–8				3–5			20–35	10–20	1–2	2–5	3–5		

Clavibacter and Curtobacterium: These have very similar profiles with very few fatty acids, most of which are branched acids. They do not contain hydroxy acids. The plant pathogenic curtobacteria can be differentiated from the *Clavibacter michiganensis* group by the absence of 15:1 *anteiso* A. *Curtobacterium flaccumfaciens* pathovars can be differentiated but there is overlap between profiles of some *Clavibacter michiganensis* subspecies.

Rhodococcus: This has a characteristic profile and contains an acid unique to plant pathogenic bacteria — 10 methyl 18:0 (tuberculostearic acid).

Computer-assisted identification

Because most genera, species, subspecies and some infraspecific taxa have unique, discrete profiles many bacteria may be rapidly and accurately identified. This can be assisted and automated by using computers. Two approaches are the use of total pattern recognition systems such as that available from Microbial ID (see above) or decision tree type systems. Both have their advantages and disadvantages. Pattern recognition systems cannot weight important acids, e.g. the hydroxy acids, and so, when present in small but extremely significant amounts, their value can be overlooked. They do however allow rapid identification of a wide range of unknown strains by reference to a database. Decision tree matrices involve more analysis but may allow differentiation of closely related bacteria which would not be so well differentiated by overall pattern recognition systems. They are perhaps of greatest use when trying to identify specific target organisms, e.g. biovars of *Agrobacterium*, soft rot *Erwinia* spp., *Xanthomonas campestris* and *Pseudomonas syringae* pathovars, from hosts such as brassicas which may have more than one pathogen.

Appendix 4: Identification of bacterial cultures by SDS-PAGE

Equipment
Electrophoresis

- Vertical electrophoresis tank (Hoefer SE600, LKB Pharmacia 16 cm vertical electrophoresis unit, Bio-Rad Protean II) with facilities for casting gels.
- Power pack to provide current.
- Cooling unit with pump to control temperature of electrophoresis buffers.
- Magnetic stirrer.
- Destaining tank with shaker. This is not essential but would be useful
- Gel dryer to dry gels for easier handling and storage.

Analysis

- Densitometer to integrate banding patterns into a trace (LKB Pharmacia Ultroscan, Joyce Loebl Chromoscan, Bio-Rad Video densitometer).
- Microcomputer, software and printer. These are used in conjunction with the densitometer.

Materials
SDS-PAGE

- Acrylamide monomer solution:

Acrylamide monomer	58.4 g
N,N'-methylene-bis-acrylamide	1.6 g

  Make up to 200 ml with distilled water (store at 4°C in the dark).
- Separating (or resolving) gel buffer: 1.5 M Tris-HCl, pH 8.8.

Tris	36.3 g

  Adjust to pH 8.8 with HCl.
  Make up to 200 ml with distilled water.
- Stacking gel buffer: 0.5 M Tris-HCl, pH 6.8.

Tris	3.0 g

  Adjust to pH 6.8 with HCl.
  Make up to 50 ml with distilled water.
- 10% SDS   500 ml
- Separating gel overlay:

Separating gel buffer	25 ml
10% SDS	1 ml

  Make up to 100 ml with distilled water.
- Sample treatment buffer:

Stacking gel buffer	2.5 ml
10% SDS	4 ml
Glycerol	2 ml
2-mercaptoethanol	1 ml

  Make up to 10 ml with distilled water and store frozen.
- Tank buffer:

Tris	12 g
Glycine	57.6 g
10% SDS	40 ml

  Make up to 4 litres with distilled water.
- Stain solution:

1% Coomassie Blue R-250	62.5 ml
Methanol	250 ml
Acetic acid	50 ml

  Make up to 500 ml with distilled water.
- Destaining solution A:

Methanol	500 ml
Acetic acid	100 ml

  Make up to 1 litre with distilled water.
- Destaining solution B:

Acetic acid	700 ml
Methanol	500 ml

  Make up to 10 litres with distilled water.

Methods
Harvesting proteins
from disrupted cells

Transfer 100 mg fresh weight of cells from culture into a 1.5 ml Eppendorf tube. Suspend in 1 ml sterile distilled water and spin for 10 min at 12 000 g to pellet the cells. Resuspend in 1 ml of sample treatment buffer and heat at 100°C for 10 min. Spin at 12 000 g for 10 min to remove cell debris. Store at −20°C until required.

Preparation of separating gel

- Assemble the vertical slab gel unit in its casting mode using the appropriate spacers (e.g. 1.5 mm spacers).
- For six 1.5 mm thick gels, mix in a 500 ml conical side-arm flask:

Acrylamide monomer solution	67 ml
Separating gel buffer	50 ml

Make up to 193 ml with distilled water.
- Attach to a vacuum pump to de-aerate for 10 min with stirring.
- Add the following.

Ammonium persulphate solution (18 mg ml^{-1})	5 ml
Distilled water	90 ml
10% SDS	2 ml
TEMED	60 µl

- Mix and fill the gel cases using a 50 ml syringe. Fill to a level about 5 cm from the top.
- Add a PVC blank comb across the top of the gel. Remove the comb after 30 min and flush the top of the gel with distilled water.
- Overlay this surface with fourfold diluted separating gel buffer and leave overnight to complete polymerization.

Preparation of stacking gel

- Mix the stacking gel as follows in a 500 ml conical side-arm flask.

Acrylamide monomer solution	10 ml
Stacking gel buffer	15 ml

Make up to 54 ml with distilled water.
- De-aerate as before and then add:

Ammonium persulphate solution (12 mg ml^{-1})	5 ml
Distilled water	60 ml
10% SDS	0.6 ml
TEMED	40 µl

- Mix the stacking gel solution and add to the top of the separating gel after draining the buffer from the surface of the gel; then insert a well-forming comb. Do not trap air bubbles. Allow to set for 30 min. Remove the comb and rinse the wells with distilled water before filling them with the tank buffer.

Loading the proteins and running the gel

Avoid using the outer 2 cm of the gel to lessen the possibility of band distortion. Each gel should have two controls; one a reference strain profile, and one a mixture of pure proteins as molecular weight markers (Pharmacia, $14-97 \times 10^3$). The latter should be dissolved in buffer containing bromophenol blue to a final concentration of 0.001%. The controls allow determination of reproducibility and subsequent correction for gel-to-gel variation.
- Apply samples (50 µl) to individual wells. Put the upper buffer chamber in place and assemble the tank.
- Fill the lower buffer chamber with tank buffer until the gel case is fully immersed. Remove any air bubbles at the base of the gel case with a Pasteur pipette.
- Add a magnetic spin bar to the lower chamber and place the whole unit on a magnetic stirrer. Pipette a drop of 0.1% phenol red solution (tracking dye) into the upper chamber and fill with tank buffer without disturbing the wells.

- Adjust the temperature of the tank buffer to 4°C.
- Replace the lid and connect to the power pack with the cathode connected to the upper chamber.
- Switch on and adjust the current to the recommended rate, e.g. 30 mA per 1.5 mm gel. The run will take 4−5 h.

Some workers prefer to use a lower current for a longer time, e.g. overnight, as it may fit in better with other work.

Staining and destaining

- Remove gel from the gel case, place in the stain solution and shake gently for least 4 h.
- Remove gel and place in destaining solution A for 1 h, again with gentle shaking.
- Transfer to destaining solution B and shake briefly.
- Dry gel between cellophane sheets on a gel dryer.

Densitometric analysis

Densitometry allows the conversion of a band pattern in the gel into an integrated trace, in which peak area is correlated to band density. When linked to a computer such traces can be used to develop a classification or similarity matrix from which identifications can be made according to a percentage similarity in the trace of an unknown with either cluster group means or individual traces. Custom-made software for this purpose is not currently available from any of the makers of densitometers (e.g. Joyce Loebl, LKB Pharmacia, Bio-Rad or Hoefer but is available from P.J.H. Jackman, 44 Havelock Road, Norwich NR3 3HG, U.K.). The software packages available with most densitometers allow pattern correction for gel-to-gel variation. Methods for calculation of similarity between patterns, similarity matrix formation, clustering and identification by comparison with libraries are reviewed by Jackman (1987). These corrections and calculations take time and perhaps discourage workers from using the technique for routine identification of large numbers of samples.

Section 2
Nucleic Acid Techniques

6 Application of double-stranded RNA analysis of plants to detect viruses, virus-like agents, virus satellites and subgenomic viral RNAs

A.T. JONES

Introduction

It is commonly believed that most, if not all, plant ribonucleic acid (RNA) is transcribed from deoxyribonucleic acid (DNA), so that normal plants should contain few, if any, double-stranded RNA (dsRNA) molecules. In practice, however, some low molecular weight ($<1.0 \times 10^6$) dsRNA often occurs in extracts from apparently healthy plants; the reason(s) for this are not clear. Nevertheless, detection of high molecular weight ($>1.0 \times 10^6$) dsRNA in plants would seem to indicate some form of abnormality, the most likely being infection by an agent with an RNA genome. A few plant viruses have dsRNA genomes and will be readily detected by dsRNA analysis of plants, but most have single-stranded RNA (ssRNA) genomes. These ssRNA viruses replicate in plants by the production of an RNA strand complementary to that of its genome. RNA strands with complementary sequences then form a base-paired double-stranded template for transcription of genomic RNA. By definition, therefore, the complementary strands should be of equal size to the genomic RNA from which they were produced, and such dsRNA structures are termed the replicative form (RF). However, at any one time during virus replication in cells, various levels of double-strandedness will occur and, on extraction, the ssRNA arms of such intermediate dsRNA forms (RI) will usually be destroyed by ribonucleases to produce some dsRNA that is smaller than that of RF. Nevertheless, the majority of dsRNA produced is likely to be RF.

Detection of dsRNA in plants is dependent on: (i) separating it from viral and host ssRNA and DNA; and (ii) detecting the very small quantities of dsRNA present. Advances in techniques for the simple separation and staining of very small quantities of nucleic acids have allowed dsRNA analysis of plant extracts to become widely available to assess infection in plants with dsRNA viruses, ssRNA viruses, virus-like agents, virus satellites and subgenomic viral RNA species. In addition, dsRNA analysis has been used to detect virus infection in fungi (Castanho et al., 1978; Morris & Dodds, 1979; Tooley et al., 1989) but this will not be considered here.

Methodology

Various procedures have been used for the separation and purification of dsRNA from plant tissue. Those most commonly used are based on the different responses of nucleic acids to either: (i) enzymatic treatments, e.g. with DNase and RNase (Breyel *et al.*, 1988); (ii) solubility in salt solutions such as LiCl (Diaz-Ruiz & Kaper, 1978); (iii) sedimentation rates in sucrose density gradients (Condit & Fraenkel-Conrat, 1979); (iv) buoyant density in Cs_2SO_4 (Bozarth, 1976); (v) molecular sieving using various media such as hydroxyapatite (Kalmakoff & Payne, 1973); or (vi) binding capacities to cellulose powder at different ethanol concentrations (Franklin, 1966; Morris & Dodds, 1979). However, most protocols tend to use combinations of these methods. Because of their simplicity, suitability for small quantities of plant material and for processing many samples simultaneously, various forms of cellulose chromotography have been widely used (Dodds *et al.*, 1984). Other benefits are that only simple and inexpensive laboratory equipment is required and that little modification is needed for different plant species, plant tissues or agents. The precise protocol used by different workers varies but most are modifications of a method described by Morris & Dodds (1979) and involve the following.

1 Disruption of plant tissue to release cell contents, usually by powdering in liquid nitrogen and/or homogenizing.

2 A buffer system to denature proteins and minimize ribonuclease activity (high pH, high salt buffer containing detergent, anti-oxidant and chelating agent).

3 Separation of nucleic acids from proteins and other cell components, usually by extraction of the buffer extract with phenol and chloroform, followed by centrifugation.

4 Fractionation of nucleic acids using cellulose powder.

5 Removal of any residual traces of DNA and ssRNA by digestion with appropriate enzymes.

6 Separation of dsRNA species by electrophoresis of samples in agarose or polyacrylamide gels and detection by staining with ethidium bromide and/or silver nitrate.

Some protocols which have been used successfully for several different viruses and plant species are given in Appendix 1.

Use of dsRNA analysis to detect viruses and virus-like agents in plants

Detection of viruses which have particles containing dsRNA

Only a few plant viruses are known to have dsRNA genomes. These viruses are of two main kinds: cryptoviruses (Boccardo *et al.*, 1987) and reoviruses (Boccardo & Milne, 1984). In plants, dsRNA produced as a result of infection with these viruses is the end product of virus replication and not a transitory product as is the situation with ssRNA viruses. Consequently, the concentration of dsRNA found in plants infected with these viruses is usually much greater than that found after infection with ssRNA viruses, and dsRNA detection is correspondingly easier.

The cryptoviruses have only recently been recognized, despite the fact that such viruses seem to be common in a wide range of plant genera. However, the very low concentration of particles of these viruses in plants, the lack of transmissibility of the viruses except through pollen and seed, and the apparently benign effects of the viruses on plants have made study difficult (Boccardo *et al.*, 1987). In addition to their intrinsic interest as viruses, they pose a problem when present in plants used to propagate other viruses by contaminating virus preparations and by inducing antibodies when such preparations are injected into animals for the production of antiserum. Purification of the particles of such viruses, although possible, is difficult, time-consuming and unsuitable for screening for infection. Abou-Elnasr *et al.* (1985) were amongst the first to use dsRNA analysis of plant tissue to detect cryptovirus infection. They found that the dsRNA of vicia cryptic virus (VCV) or similar viruses, was readily detected in *Vicia faba* plants using as little as one leaflet (0.6 g) and the technique proved sensitive enough to detect VCV in extracts from *V. faba* protoplasts (Fig. 6.1). Many plants were screened in this way and VCV infection was found to be widespread in *V. faba* seed stocks and to occur in 18 of 20 cultivars tested (Fig. 6.1). The technique was also used to detect cryptovirus-like dsRNA in leaves and protoplasts of brassicas (Jones *et al.*, 1986 a) and in leaves of cucumber (*Cucumis sativus*) (Nameth & Dodds, 1985; Jelkmann *et al.*, 1988).

Detection of viruses which have particles containing ssRNA

Considerably less dsRNA is found in plants infected with viruses having ssRNA genomes, than with those having dsRNA genomes. In many instances, however, this is compensated somewhat by the high virus titre that occurs in plants infected with some ssRNA viruses. Indeed, dsRNA was readily detected in plants infected with viruses representing most of the groups that have ssRNA genomes, including bromo-, carmo-, como-, cucumo-, diantho-, ilar-, necro-, nepo-, sobemo-, tombus-, capillo-, carla-, clostro-, furo-, potex-, poty-, tobamo- and tobraviruses (Dodds, 1986; Jones *et al.*, 1986 b; Valverde *et al.*, 1986; Rochon & Tremaine, 1988; Jones & Mitchell, 1988; A.T. Jones & W. McGavin, unpublished results). For most of the viruses in these groups, dsRNA analysis would not usually be the detection method of choice as the viruses are readily transmitted by

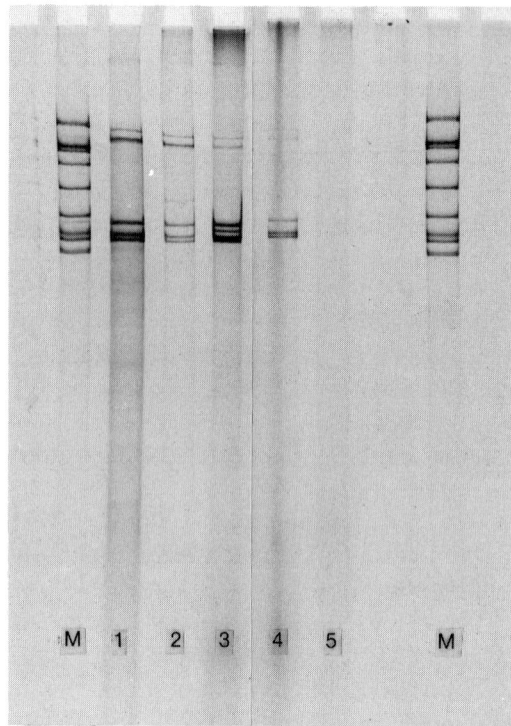

Fig. 6.1 Polyacrylamide gel electrophoresis of dsRNA extracted from $1-2$ g leaf tissue from five *Vicia faba* cultivars (lanes $1-5$) and from purified particles of maize rough dwarf reovirus used as markers (mol. wt range $2.85 \times 10^6 - 1.13 \times 10^6$, lane M). Vicia cryptic virus-like dsRNA is present in four (lanes $1-4$) of the five cultivars.

inoculation of plant sap, reach large concentrations in plants and can be detected and studied more readily in other ways. Nevertheless, even for some of these viruses, dsRNA analysis has proved useful to compare and identify different virus strains, for example those of citrus tristeza (CTV; Dodds *et al.*, 1985 b, 1987), tobacco mosaic (TMV; Valverde *et al.*, 1986) and cucumber mosaic (CMV; Dodds *et al.*, 1985 a; Wang *et al.*, 1988) viruses. Strain identification by this means provided a useful marker in a study of cross-protection between strains of CMV (Dodds *et al.*, 1985 a). Furthermore, comparison of the dsRNA profiles from plants infected with cucumber necrosis (CNV), tomato bushy stunt (TBSV) and tobacco necrosis (TNV) viruses, taken together with other data, clearly distinguished CNV from TNV and showed its affinity to TBSV and other tombusviruses (Rochon & Tremaine, 1988).

However, dsRNA analysis has proved to be especially useful for detecting, studying and characterizing some viruses that are less amenable to study by conventional means because of the very low concentration of their particles in plants and/or the instability of their particles, and/or their poor transmissibility between plants by mechanical inoculation of sap extracts. Thus, dsRNA analysis together with other biological data has provided evidence to suggest a luteovirus as the possible causal agent of banana bunchy top disease (Dale *et al.*, 1986) and of strawberry mild yellow edge disease (Spiegel, 1987), and different dsRNA profiles were associated with different strains of barley yellow dwarf (Gildow *et al.*, 1983) and beet western yellows (Falk & Duffus, 1984) luteoviruses. However, to detect the dsRNA species associated with these luteo- or luteo-like viruses much larger amounts of leaves (up to 100 g) are required than with viruses that tend to reach much higher concentrations in plants.

Plants infected with carrot mottle, groundnut rosette and lettuce speckles mottle viruses, for which no particles have been identified with confidence, have yielded several dsRNA species allowing detection and comparison of these viruses (Fig. 6.2); such studies have strengthened the conclusions drawn from other data in indicating that these viruses are very similar (Reddy *et al.*, 1985; Falk *et al.*, 1979).

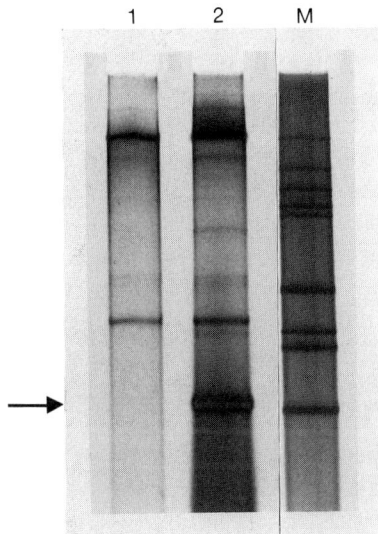

Fig. 6.2 Polyacrylamide gel electrophoresis of dsRNA extracted from *Nicotiana benthamiana* leaves infected with a satellite-free (lane 1) and satellite-containing (arrow, lane 2) culture of groundnut rosette virus. Lane M contains dsRNA from rice dwarf reovirus used as mol. wt markers (range $3.1 \times 10^6 - 0.48 \times 10^6$).

Analysis of dsRNA in plants has also aided the study of other 'difficult' viruses or virus strains, for example, strawberry mottle (Adams & Barbara, 1986), an avirulent strain of plum pox (Breyel *et al.*, 1986), an aphid-transmitted virus of *Rubus* in North America (Kurppa & Martin, 1986; Jelkmann & Martin, 1989) and a possible mealybug-transmitted virus of pineapple (Gunasinghe & German, 1988).

Detection of agents of diseases of presumed viral aetiology

Possibly one of the greatest benefits of current dsRNA technology has been its usefulness in detecting viruses or virus-like agents of diseases that have proved intractable to study using other virological techniques. The agent of lettuce big vein disease, transmitted by the fungus *Olpidium brassicae*, has proved difficult to study for very many years but Mirkov and Dodds (1985) reported several dsRNA species specifically associated with extracts from roots of affected plants.

June yellows of strawberry is another condition that has remained an enigma for more than 100 years. The condition is seed- and pollen-borne but is not transmissible by grafting, and was thought to be a genetic disorder (Hughes, 1989). However, recent studies have associated small quantities of several high molecular weight dsRNA species with June yellows (Watkins *et al.*, 1990). Although it is not known if these dsRNA species are derived from a causal agent or are a consequence of June yellows, dsRNA analysis has gone some way towards providing an objective test for the condition.

Studies for many years on several virus-like diseases of black currant have been hindered by the intractable nature of the agents involved. However, plants affected by some of these diseases, and some apparently healthy plants of several older cultivars, contain distinct dsRNA species not present in more recently introduced cultivars (Fig. 6.3). These dsRNA species can be classified in one or other of about four groups (Jones *et al.*, 1986 b; A.T. Jones & K. Knoll, unpublished results). This information provides a promising lead to understanding and studying the agents involved in these diseases. Similarly, studies by Di *et al.* (1989) on rose rosette, a disease of unknown aetiology affecting *Rosa multiflora* in North America, identified several dsRNA species associated with this condition.

Analysis of dsRNA in plants has also provided evidence of infection with agents having RNA genomes in crop plants with no obvious symptoms, e.g. avocado (Jordan *et al.*, 1983). Undoubtedly the technique will continue to prove very valuable in studying the virus pathology of plants.

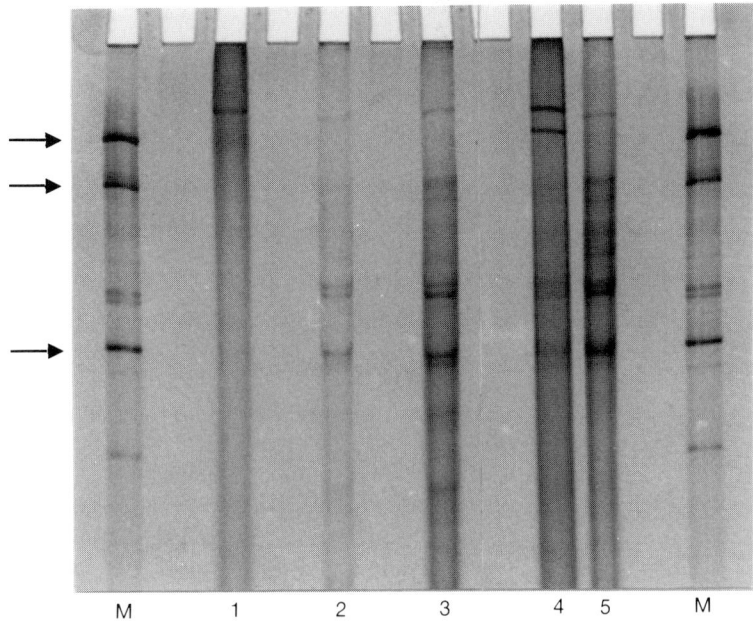

Fig. 6.3 Polyacrylamide gel electrophoresis of dsRNA extracted from leaves of five apparently healthy black currant cultivars (lanes 1−5) and from *Nicotiana clevelandii* leaves infected with cucumber mosaic virus used as a marker (lanes M, arrows indicate mol. wts of 2.57×10^6, 1.76×10^6 and 0.46×10^6). All five cultivars contain one or more high mol. wt dsRNA species.

Use of dsRNA analysis to detect virus satellite RNA and viral subgenomic RNA

Detection of virus satellite RNA in plants

The RNA of virus satellites often reaches a much greater concentration in plants than that of the RNA of the helper virus and, not surprisingly, dsRNA analysis of several virus satellite systems has shown a similar difference in the concentrations of their respective RF dsRNA species. Thus, satellites of CMV are readily detected in plants by dsRNA analysis (Morris *et al.*, 1983; White & Kaper, 1989), and analysis of dsRNA was used to demonstrate the presence of a satellite RNA in a strain of TMV and to assess its frequency in the field (Valverde & Dodds, 1986). A satellite RNA of groundnut rosette virus was also readily detected using dsRNA analysis and shown to be the main cause of rosette disease in groundnut (Fig. 6.2; Murant *et al.*, 1988) and a simple and rapid diagnostic test for this dsRNA species was proposed (Breyel *et al.*, 1988).

Detection of viral subgenomic RNA in plants

In addition to RF dsRNA that is twice the size of genomic viral RNA, dsRNA analysis of virus-infected plants often yields smaller amounts of lower molecular weight dsRNA not associated with any virus satellite RNA. Such minor dsRNA species are commonly detected in a wide range of virus–host combinations and it has been suggested (Dawson & Dodds, 1982) that they are involved in the synthesis of subgenomic mRNA. Although their precise role is obscure, several of these minor dsRNA species have been shown to be the double-stranded form of viral subgenomic RNA. Thus, Dawson and Dodds (1982) found seven of 12 dsRNA species in TMV-infected plants to correspond in size to what would be expected for the RF form of the subgenomic RNA species reported for TMV and that one, which was especially abundant, corresponded in size to TMV coat protein mRNA. Studies of dsRNA from cucumovirus (Kaper & Diaz-Ruiz, 1977) and potexvirus (Guilford & Forster, 1986; Mackie *et al.*, 1988) infections in plants have identified what are also believed to be subgenomic RNA species including those reported to be the coat protein mRNA. In extracts of plants infected with CTV, Dulieu & Bar-Joseph (1990) not only identified viral subgenomic dsRNA species but also, by *in vitro* translation studies, showed that the smallest species is the RF of the coat protein mRNA. Thus, analysis of dsRNA from plants infected with some viruses, particularly those like closteroviruses that have a very large genome, can be useful for studying virus genome strategies.

Precautions in dsRNA analysis of plant material

The use of dsRNA analysis is, and will continue to be, a useful technique for detection of viruses, virus-like agents and their associated RNA species in plants. However, caution is necessary in the assessment and interpretation of results from such analyses because of the sensitivity of the detection system and the risks of contamination. Most of the examples of the use of dsRNA analysis of plant material given above are from instances where plants are infected with a single virus or agent. However, multiple infections are common in the field, and may also occur in laboratory conditions, making the interpretation of dsRNA profiles difficult. Particular care should be given to assessing the risk of contamination by cryptoviruses even in plants grown from seed and, where possible, all dsRNA assays should include directly comparable controls. Contamination can arise from sources other than plant tissue itself. Virus infection is common in arthropods that infest plants, including those that are considered not to be vectors of plant viruses. A few such infected individuals, or even their carcasses, can be a potent source of contamination. For example, Watkins *et al.* (1990) found significant quantities of about 10 dsRNA species in extracts of the two-spotted spider mite, *Tetranychus urticae*, colonizing plants in the glasshouse and in an extract of a single leafhopper collected from the field.

Finally, the occurrence of high molecular weight dsRNA species in plants, though indicative of some abnormality, may not necessarily be indicative of infection with a virus or virus-like agent. Some dsRNA species detected in *Vicia faba* (Grill & Garger, 1981), *Phaseolus vulgaris* (Wakarchuk & Hamilton, 1985) and in *Rubus* spp. (Stace-Smith & Martin, 1989) appear to be produced by transcription from plant DNA because cDNA probes made from these dsRNA species hybridized to DNA extracted from normal-looking plants.

Other uses of dsRNA analysis

In many instances, the analysis of dsRNA can form the basis of further studies. These include the following.

Infectivity assays: Several hosts, particularly woody hosts, contain chemicals which interfere with mechanical transmission of some viruses. One way of overcoming this is to use nucleic acids from such plants as inoculum. As dsRNA is much less sensitive to ribonuclease than ssRNA, unfractionated dsRNA extracts or gel-purified dsRNA can be prepared, then melted to separate the RNA strands and used to inoculate plants.

Preparation of hybridization probes: A band containing a selected species of dsRNA can be excised from gels, the dsRNA species extracted, denatured to separate the strands and the ssRNA labelled at the 5′ end with polynucleotide kinase to produce a probe for hybridization tests (Jordan & Dodds, 1983; Kurppa & Martin, 1986). Alternatively, the gel-purified dsRNA species can be denatured before use as a template to produce a cDNA probe by reverse transcription (Maniatis *et al.*, 1982; Jelkmann & Martin, 1989). The production of probes in these ways allows the specific and sensitive detection of the virus or agent involved without the need to purify its particles (Jelkmann & Martin, 1989). Using this approach an agent detected in pineapple with pineapple wilt symptoms was also detected in mealybugs (Gunasinghe & German, 1988).

In vitro *translation from dsRNA species*: Gel-purified dsRNA can be denatured and used directly for translation studies (Dulieu & Bar-Joseph, 1990). If a viral (agent) protein can be produced in this way from mRNA, it may be possible to produce an antiserum, specific to the virus (agent), without the need to purify its particles.

Conclusions

Analysis of dsRNA from plants has proved a useful and versatile technique for detecting and, in some instances, studying plant viruses, virus-like agents, subgenomic RNAs and satellite RNAs. As already indicated, this technique has been especially useful for detecting agents that have been difficult to study in other ways and, for some such agents, it remains the only technique currently available. Nevertheless, relatively few examples exist of the use of purified dsRNA for other studies such as *in vitro* translation and as a template to produce nucleic acid probes. The extension of dsRNA technology into these areas promises to increase further the usefulness of this novel approach to study RNA-containing viruses and virus-like agents.

References

Abou-Elnasr M.A., Jones A.T. & Mayo M.A. (1985) Detection of dsRNA in particles of vicia cryptic virus and in *Vicia faba* tissues and protoplasts. *Journal of General Virology* **66**, 2453−60.

Adams A.N. & Barbara D.J. (1986) Transmission of a virus from *Fragaria vesca* infected with strawberry mottle virus to *Chenopodium quinoa*. *Acta Horticulturae* **186**, 71−6.

Boccardo G. & Milne R.G. (1984) Plant reovirus group. In Murant A.F. & Harrison B.D. (eds) *Commonwealth Mycological Institute/Association of Applied Biologists Descriptions of Plant Viruses*, No. 294, 7 pp. Association of Applied Biologists, Wellesbourne.

Boccardo G., Lisa V., Luisoni E. & Milne R.G. (1987) Cryptic plant viruses. *Advances in Virus Research* **32**, 171−214.

Bozarth R.F. (1976) The buoyant density of three double-stranded RNAs in cesium sulfate. *Biochemica Biophysica Acta* **442**, 32−6.

Breyel E., Maiss E., Casper R. & El-Ouaghlidi F. (1986) Isolation of dsRNA from plum pox virus-infected plant tissue. *Acta Horticulturae* **193**, 167−72.

Breyel E., Casper R., Ansa O.A., Kuhn C.W., Misari S.M. & Demski J.W. (1988) A simple procedure to detect a dsRNA associated with groundnut rosette. *Journal of Phytopathology* **121**, 118−24.

Castanho B., Butler E.E. & Shepherd R.J. (1978) The association of double-stranded RNA with *Rhizoctonia* decline. *Phytopathology* **68**, 1515−19.

Condit C. & Fraenkel-Conrat H. (1979) Isolation of replicative forms of 3′ terminal subgenomic RNAs of tobacco necrosis virus. *Virology* **97**, 122−30.

Dale J.L., Phillips D.A. & Parry J.N. (1986) Double-stranded RNA in banana plants with bunchy top disease. *Journal of General Virology* **67**, 371−5.

Dawson W.O. & Dodds J.A. (1982) Characterization of sub-genomic double-stranded RNAs from virus-infected plants. *Biochemical and Biophysical Research Communications* **107**, 1230−5.

Di R., Epstein A.H. & Hill J.H. (1989) Double-stranded RNA associated with multiflora rose expressing symptoms of rose rosette disease. *Phytopathology* **79**, 1139.

Diaz-Ruiz J.R. & Kaper J.M. (1978) Isolation of viral double-stranded RNAs using LiCl fractionation procedure. *Preparative Biochemistry* **8**, 1−17.

Dodds J.A. (1986) The potential for using double-stranded RNAs as diagnostic probes for plant viruses. In Jones R.A.C. & Torrance L. (eds) *Developments and Applications in Virus Testing*, pp. 71−86. Association of Applied Biologists, Wellesbourne.

Dodds J.A., Morris T.J. & Jordan R.L. (1984) Plant viral double-stranded RNA. *Annual Review of Phytopathology* **22**, 151−68.

Dodds J.A., Lee S.Q. & Tiffany M. (1985 a) Cross protection between strains of cucumber mosaic virus: effect of host and type of inoculum on accumulation of virions and double-stranded RNA of the challenge strain. *Virology* **144**, 301–9.

Dodds J.A, Tamaki S.J. & Roistacher C.N. (1985 b) Indexing of citrus tristeza virus double-stranded RNA in field trees. In Garnsey S.M., Timmer L.W. & Dodds J.A. (eds) *Proceedings of Ninth Conference of the IOCV 1984*, pp. 327–9. International Organization of Citrus Virologists, University of California, Riverside.

Dodds J.A., Jordan R.L., Roistacher C.N. & Jarupat T. (1987) Diversity of citrus tristeza virus isolates indicated by dsRNA analysis. *Intervirology* **27**, 177–88.

Dulieu P. & Bar-Joseph M. (1990) *In vitro* translation of the citrus tristeza virus coat protein from an 0.8 kbp double-stranded RNA segment. *Journal of General Virology* **71**, 443–7.

Falk B.W. & Duffus J.E. (1984) Identification of small single- and double-stranded RNAs associated with severe symptoms in beet western yellows virus-infected *Capsella bursa-pastoris*. *Phytopathology* **74**, 1224–9.

Falk B.W., Morris T.J. & Duffus J.E. (1979) Unstable infectivity and sediment-able dsRNA associated with lettuce speckles mottle virus. *Virology* **96**, 239–48.

Franklin R.M. (1966) Purification and properties of replicative intermediate of the RNA bacteriophage R 17. *Proceedings of the National Academy of Science USA* **55**, 1504–11.

Gildow F.E., Ballinger M.E. & Rochow W.F. (1983) Identification of double-stranded RNAs associated with barley yellow dwarf virus infection in oats. *Phytopathology* **73**, 1570–2.

Grill L.K. & Garger S.J. (1981) Identification and characterization of double-stranded RNA associated with cytoplasmic male sterility in *Vicia faba*. *Proceedings of the National Academy of Science USA* **78**, 7043–6.

Guilford P.J. & Forster R.L.S. (1986) Detection of polyadenylated subgenomic RNAs in leaves infected with the potexvirus daphne virus X. *Journal of General Virology* **67**, 83–90.

Gunasinghe U.B. & German T.L. (1988) Detection of viral RNA in mealybugs associated with mealybug-wilt of pineapple. *Phytopathology* **78**, 1584.

Hughes J. d'A. (1989) Strawberry June yellows – a review. *Plant Pathology* **38**, 146–60.

Jelkmann W. & Martin R.R. (1989) Complementary DNA probes generated from double-stranded RNAs of a *Rubus* virus provide the potential for rapid *in vitro* detection. *Acta Horticulturae* **236**, 103–9.

Jelkmann W., Maiss E., Casper R. & Lesemann D.E. (1988) *Cucumis sativus* cryptic virus, a new virus in cucumber. *Journal of Phytopathology* **121**, 233–8.

Jordan R. & Dodds J.A. (1983) Hybridization of 5'-end-labelled RNA to plant viral RNA in agarose and acrylamide gels. *Plant Molecular Biology Reporter* **1**, 31–7.

Jordan R., Dodds J.A. & Ohr H. (1983) Evidence for virus-like agents in avocado. *Phytopathology* **73**, 1130–5.

Jones A.T. & Mitchell M.J. (1988) Further studies on double-stranded RNA species from raspberry plants infected with aphid-borne viruses. *Annals of Applied Biology* **113**, 431–6.

Jones A.T., Abou-Elnasr M.A., Mayo M.A. & Hodgkin J.R.T. (1986 a) Detection of viruses in brassicas. *Report of the Scottish Crop Research Institute for 1985*, pp. 169–70. Scottish Crop Research Institute, Invergowrie, Dundee.

Jones A.T., Abou-Elnasr M.A., Mayo M.A. & Mitchell M.J. (1986 b) Association of dsRNA species with some virus-like diseases of small fruits. *Acta Horticulturae* **186**, 63–70.

Kalmakoff J. & Payne C.C. (1973) A simple method for the separation of single- and double-stranded RNA on hydroxyapatite. *Annals of Biochemistry* **55**, 26–33.

Kaper J.M. & Diaz-Ruiz J.R. (1977) Molecular weights of the double-stranded RNAs of cucumber mosaic virus strain S and its associated RNA5. *Virology* **80**, 214–17.

Kurppa A. & Martin R.R. (1986) Use of double-stranded RNA for detection and identification of virus diseases of *Rubus* species. *Acta Horticulturae* **186**, 51–62.

Mackie G.A., Johnston R. & Bancroft J.B. (1988) Single- and double-stranded viral RNAs in plants infected with the potexviruses papaya mosaic virus and foxtail mosaic virus. *Intervirology* **29**, 170–7.

Maniatis T., Fritsch E.F. & Sambrook J. (1982) *Molecular Cloning: A Laboratory Manual*, pp. 187–211. Cold Spring Harbor Laboratory, Cold Spring Harbor, New York.

Mirkov T.E. & Dodds J.A. (1985) Association of double-stranded ribonucleic acids with lettuce big vein disease. *Phytopathology* **75**, 631–5.

Morris T.J. & Dodds J.A. (1979) Isolation and analysis of double-stranded RNA from virus-infected plant and fungal tissue. *Phytopathology* **69**, 854–8.

Morris T.J., Dodds J.A., Hillman B., Jordan R.L., Lommel S.A. & Tamaki S.J. (1983) Viral specific dsRNA: diagnostic value for plant virus disease identification. *Plant Molecular Biology Reporter* **1**, 27–30.

Murant A.F., Rajeshwari R., Robinson D.J. & Raschke J.H. (1988) A satellite RNA of groundnut rosette virus that is largely responsible for symptoms of groundnut rosette disease. *Journal of General Virology* **69**, 1479–86.

Nameth S.T. & Dodds J.A. (1985) Double-stranded RNAs detected in cucurbit varieties not inoculated with viruses. *Phytopathology* **75**, 1293.

Reddy D.V.R., Murant A.F., Raschke J.H., Mayo M.A. & Ansa O.A. (1985) Properties and partial purification of infective material from plants containing groundnut rosette virus. *Annals of Applied Biology* **107**, 65–78.

Rochon d'A. & Tremaine J.H. (1988) Cucumber necrosis virus is a member of the tombusvirus group. *Journal of General Virology* **69**, 395–400.

Spiegel S. (1987) Double-stranded RNA in strawberry plants infected with strawberry mild yellow-edge virus. *Phytopathology* **77**, 1492–4.

Stace-Smith R. & Martin R.R. (1989) Occurrence of seed-transmissible double-stranded RNA in native red and black raspberry. *Acta Horticulturae* **236**, 13–20.

Tooley P.W., Hewings A.D. & Falkenstein K.F. (1989) Detection of double-stranded RNA in *Phytophthora infestans*. *Phytopathology* **79**, 470–4.

Valverde R.A. & Dodds J.A. (1986) Evidence for a satellite RNA associated naturally with the U5 strain and experimentally with the U1 strain of tobacco mosaic virus. *Journal of General Virology* **67**, 1875–84.

Valverde R.A., Dodds J.A. & Heick J.A. (1986) Double-stranded ribonucleic acid from plants infected with viruses having elongated particles and undivided genomes. *Phytopathology* **76**, 459–65.

Wakarchuk D.A. & Hamilton R.I. (1985) Cellular double-stranded RNA in *Phaseolus vulgaris*. *Plant Molecular Biology* **5**, 55–63.

Wang W.Q., Natsuaki T., Okuda S. & Teranaka M. (1988) Comparison of cucumber mosaic virus isolates by double-stranded RNA analysis. *Annals of the Phytopathological Society of Japan* **54**, 536–9.

Watkins C.A., Jones A.T., Mayo M.A. & Mitchell M.J. (1990) Double-stranded RNA analysis of strawberry plants affected by June yellows. *Annals of Applied Biology* **117**, 73–83.

White J.L. & Kaper J.M. (1989) A simple method for detection of viral satellite RNAs in small plant tissue samples. *Journal of Virological Methods* **23**, 83–94.

Appendix 1: Protocols for extraction and detection of dsRNA from plant tissue

The following protocols have proved useful in extraction of dsRNA from a range of tissues and plant hosts. Protocol A was designed specifically for use with strawberry tissue which poses several problems not encountered with more herbaceous material (Watkins *et al.*, 1990). Protocol B is a more rapid procedure based on that of Morris *et al.* (1983) and which eliminates the DNase digestion and second cellulose chromatography step; it is suitable for less difficult material and where dsRNA concentrations are higher than they are in strawberry. For convenience, each protocol is based on the use of 10 g of starting material.

Protocol A
- Powder 10 g fresh or frozen tissue in liquid nitrogen in a pestle and mortar and extract in 20 ml TMS buffer (100 mM Tris, 10 mM magnesium acetate, 500 mM NaCl, pH 8.5) containing 2% (w/v) (mol. wt 44 000) PVP, 1% (w/v) SDS, 1% (w/v) bentonite, 0.2% (w/v) DIECA and 0.1% (w/v) thioglycerol.
- Add 20 ml water-saturated phenol containing 10% (v/v) *m*-cresol, 0.3% 8-hydroxyquinoline, and 20 ml chloroform−pentanol (25:1) and stir.
- Incubate at 60°C for 15 min.
- Centrifuge at 10 000 g for 10 min and recover the aqueous supernatant phase.
- To each 1 ml of this aqueous phase add 0.02 g cellulose powder (Cellex N-1 or Whatman CF-11) and 0.22 ml ethanol.
- Stir the suspension for 30 min at room temperature.
- Centrifuge at 10 000 g for 10 min and discard the supernatant fluid.
- Resuspend the cellulose pellet in 1−2 ml STE buffer (200 mM NaCl, 50 mM Tris, 1 mM EDTA, pH 7) containing 18% ethanol.
- Centrifuge at 10 000 g for 10 min and discard the supernatant fluid.
- Wash the cellulose pellet twice with 18% ethanol in STE buffer, recovering the cellulose pellet by centrifuging.
- During the final wash, load the cellulose suspension in 18% ethanol into a glass column or disposable syringe barrel and allow to drain.
- Elute the dsRNA and any remaining contaminating DNA from the cellulose with STE buffer without ethanol.
- Precipitate the eluted nucleic acids from the buffer by adding 2.5 vol. ethanol and storing overnight at −15°C or for a few hours at −70°C.
- Recover the nucleic acids by centrifuging at 10 000 g for 10 min, discard the supernatant fluid and resuspend the pellet in 1 ml STE buffer containing 30 mM MgCl$_2$.
- Add 10 µg ml^{-1} DNase and incubate at 30°C for 60 min.
- Add 1 ml 20 mM Tris buffer containing 1 mM EDTA and 4% (w/v) SDS and incubate at 60°C for a further 15 min.
- Add an equal volume of water-saturated phenol and emulsify.
- Centrifuge at 10 000 g for 10 min, recover the aqueous supernatant phase and add 2.5 vol. ethanol and store overnight at −15°C or for a few hours at −70°C, to precipitate the dsRNA.

- Centrifuge at 10 000 g for 10 min, discard the supernatant fluid and dry the pelleted dsRNA. Resuspend in 100 μl TPE buffer (35 mM Tris, 30 mM NaH$_2$PO$_4$, 1 mM EDTA, pH 7.6) containing 10% (w/v) RNase-free sucrose and a trace of bromophenol blue.
- Centrifuge at 5 000 g for 5 min to clarify the solution and load 20–40 μl samples per gel track on 7% polyacrylamide gels or 1% agarose in TPE buffer.
- Electrophorese at 50 V for 16–20 h for polyacrylamide gels or 3–5 h for agarose gels.
- Stain gels with ethidium bromide or silver nitrate following standard protocols.

Protocol B

- Powder 10 g of fresh or frozen tissue in liquid nitrogen in a pestle and mortar and extract the powder in 10 ml (or more) double-strength STE buffer (1× STE = 100 mM NaCl, 50 mM Tris, 1 mM EDTA, pH 8) containing 1% (w/v) SDS and 0.1% (v/v) 2-mercaptoethanol.
- Add 10 ml water-saturated phenol and 5 ml chloroform–pentanol (25:1) and stir.
- Clarify the extract by centrifuging at 10 000 g for 10 min and recover the aqueous supernatant phase.
- To this aqueous phase add ethanol to a final concentration of 16–18%.
- Pour slowly into a small column containing cellulose powder (CF-11 Whatman) and equilibrated with 18% ethanol in STE buffer.
- Wash the cellulose with several volumes of STE buffer containing 18% ethanol to elute ssRNA and DNA.
- Elute dsRNA with a small volume (1–3 ml) of STE buffer without ethanol.
- Add 2.5 vol. ethanol to the eluate and store overnight at −15°C or for a few hours at −70°C to precipitate the dsRNA.
- Recover the dsRNA by centrifuging at 10 000 g for 10 min, discard the supernatant fluid and dry the pelleted dsRNA. Resuspend in 100 μl TPE buffer containing 10% (w/v) RNase-free sucrose and a trace of bromophenol blue.
- Centrifuge at 5 000 g for 5 min to clarify the solution before electrophoresis in gels as described in Protocol A.

Problems sometimes occur with tissue containing high levels of tannins, latex or carbohydrate. Some of these problems can be overcome by increasing further the ratio of buffer volume to weight of tissue.

Marker dsRNA: It is important to have suitable high and low molecular weight dsRNA marker species to estimate the sizes of unknown dsRNA species. Workers have used dsRNA species from characterized reoviruses or mycoviruses, or dsRNA from plants infected with CMV and/or TMV.

7 Detection of viroids and viruses by nucleic acid probes

L.F. SALAZAR & M. QUERCI

Introduction

Molecular hybridization techniques for routine testing for viroids and viruses have gained worldwide acceptance in recent years. The technique involves the use of labelled complementary DNA or RNA (cDNA, cRNA) prepared from purified viroid or viral nucleic acid (or a recombinant clone of such nucleic acid) as a probe. On incubation with plant extracts these probes will detect the presence of viral or viroid nucleic acids by forming hybrids with them.

The degree of specificity of detection is determined by the degree of sequence complementarity between the RNA or DNA of the probe and the nucleic acid of the viroid or virus. Whereas the nucleotide sequence of a single isolate of a given virus or viroid is essentially constant, the sequences of different isolates of the same virus or viroid, or of unrelated viruses or viroids, can differ by as little as a few nucleotides up to a large portion of their nucleic acids. Nucleic acid probes prepared to one isolate of a viroid or virus will therefore form highly specific hybrids with the nucleic acid of that isolate, less specific ones with other isolates that differ in a few sequences and none at all with unrelated ones.

Originally nucleic acid hybridization between DNA and RNA was done in liquid (Spiegelman, 1964), but nowadays it is commonly done by blotting the test samples onto a nitrocellulose membrane and then incubating the membrane in a solution of the probe. This technique is called the nucleic acid spot hybridization (NASH) test, although other names such as dot-blot or sap-spot hybridization are also used (Boulton et al., 1984). In the NASH test, RNA or DNA can be hybridized with a RNA or DNA probe.

The NASH test was first used to detect viroids by Owens and Diener (1981). Thereafter it became the method of choice for viroids (Salazar et al., 1983), because viroids, unlike viruses, do not have an antigenic protein coat and thus cannot be detected by immunological methods (Diener, 1979). Previous methods for detecting viroids were either not sensitive enough or were inadequate for large-scale testing. For example, detection of avocado sunblotch viroid (ASBVd) by symptom development in suitable indicator hosts may take up to 2 years. Even where symptom development takes only a few weeks, the requirement for large amounts of bench space in a heated glasshouse makes bioassays unsuitable for routine indexing.

Polyacrylamide gel electrophoresis (PAGE) of naturally infected plant samples on its own is not sensitive enough to detect low levels of viroid. However, the technique can be very sensitive for potato spindle tuber viroid (PSTVd) detection if the viroid is first inoculated from potato tissue onto tomato (in which PSTVd reaches much higher concentrations) prior to using PAGE (Harris et al., 1984), but this is a cumbersome technique. Singh & Boucher (1987) have recently reported a 'return' gel electrophoresis method with a sensitivity for the detection of PSTVd matching that of NASH, but we have been unable to confirm this in tests at the International Potato Center (CIP). A possible advantage of their technique is that it might distinguish between mild and severe strains of PSTVd. NASH, on the other hand, can detect as little as 0.33 pg of PSTVd in potato plants grown under high temperature regimes, but it cannot detect the viroid in plants or tubers which have been maintained below 10°C (Salazar et al., 1988).

The NASH test has also been applied to the detection of viruses (Maule et al., 1983; Baulcombe et al., 1984 a, b). It has several advantages over serology. The latter is based on the detection of epitopes of the virus coat protein, the cistron for which represents only a small portion of the genetic information of the virus: for tobacco mosaic virus (TMV) less than 2% of the viral genome is involved in the antigenicity of the coat protein (Hull, 1986). By contrast, probes for hybridization analysis can represent the whole genome of the virus or parts thereof, which opens up new possibilities for the study of relationships between viruses. Specific probes can be used for specific purposes: cDNA clones of one virus strain have been differentially hybridized with the RNA from other strains for a number of viruses (Baulcombe et al., 1984 a; Rosner & Bar-Joseph, 1984; Gallitelli et al., 1985; Linthorst & Bol, 1986). Several probes can be combined simply by mixing them during hybridization, and can thus be used in a polyvalent manner to detect several virus strains simultaneously (Hopp et al., 1988). Single probes can be constructed which contain short, specific sequences for each of several strains of a virus or even different viruses (M. Querci & L. Salazar, unpublished results).

Despite the wide application of enzyme-linked immunosorbent assay (ELISA; Clark & Adams, 1977) to the detection of many viruses there are instances where it is inapplicable. An example is tobacco rattle virus (TRV), where infection with TRV-RNA 1 will result in replication of RNA 1 but not in the production of virions, the coat protein for which is encoded by RNA 2 (Harrison & Robinson, 1982). Under such conditions serology cannot be applied and the detection of the TRV-RNA 1 is only possible by infectivity assays of extracted nucleic acids on suitable host plants, or by NASH. Linthorst & Bol (1986) were able to develop a number of probes that could be used to detect either a wide spectrum or specific groups of TRV isolates. Moreover, NASH was reported to be more sensitive than ELISA for the detection of potato leafroll virus (PLRV) or potato virus X (PVX) in symptomless plants (Boulton et al., 1984).

One drawback to the use of probes as commonly applied is that they are radioactive. This places limitations on their use in many countries and situations. The development of non-radioactive labels for probes makes possible wider application of the technology (Leary et al., 1983; Vivian, this volume).

The NASH test, with several methods of probe preparation, has been used at CIP since 1983 for the routine detection of PSTVd (Salazar et al., 1988), and since 1987 its use has been expanded to the detection of several viruses and viroids of crops other than potato (Fig. 7.1).

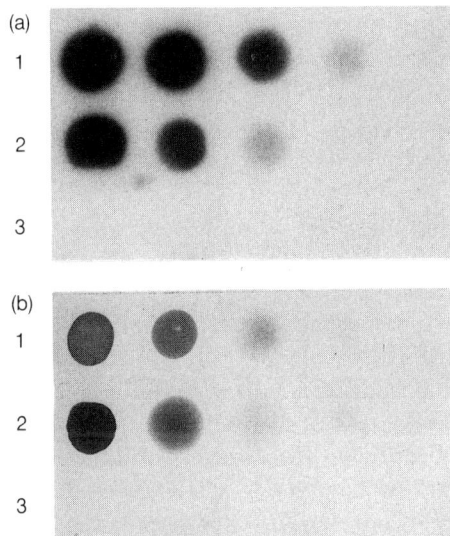

Fig. 7.1 Detection of potato virus X (PVX) by nucleic acid spot hybridization. Comparison of (a) a [^{32}P]RNA probe and (b) a biotin-labelled DNA probe. 1 = tenfold dilutions (1/10−1/10 000) of PVX infective potato sap in 2x SSC; 2 = purified PVX: 100 ng, 10 ng, 1 ng, 0.1 ng; 3 = tenfold dilutions of non-infective potato sap (1/10−1/10 000) in 2× conc. SSC.

Preparation of nucleic acid probes

Nucleic acid probes can be of two types: cDNA or cRNA probes. The probe consists of a strand of DNA or RNA, complementary to the target nucleic acid, which is conjugated to a label. The label can be a radioisotope such as ^{32}P or ^{35}S, or non-radioactive such as biotin.

Preparation of recombinant DNA clones

Despite the relative ease by which cDNA probes can be obtained from purified viroid or virus nucleic acid most are prepared using some form of recombinant DNA technology. There are several ways that this can be done and we give details of the methods used by Owens & Cress (1980) and Cress et al. (1983) to produce probes to PSTVd.

Purified PSTVd is treated with alkaline phophatase from *Escherichia coli* to remove 3'-terminal phosphate residues from linear molecules. A polyadenylate tail is added by incubation with poly(A)polymerase, and the polyadenylated form of the PSTVd is recovered by phenol:chloroform extraction and ethanol precipitation. Single-stranded cDNA of PSTVd is synthesized by incubation with reverse transcriptase and the four deoxynucleoside triphosphates (dNTPs). For synthesis of the second strand, the single-stranded cDNA is denatured at 100°C, quenched at 0°C, and added to a reaction mixture containing DNA polymerase I and the four dNTPs. The non-base-paired regions of the double-stranded DNA are removed by incubation with S1 nuclease.

Hybrids are constructed from plasmid pBR322 and S1-digested double-stranded PSTVd cDNA, and annealed before transformation of *E. coli* C600 (rk$^-$ mk$^+$). Tetracycline-resistant and ampicillin-sensitive transformants are screened, the cloned DNAs are isolated and their size and reactivity assessed. Two clones obtained in this way, pDC-29 and pDC-22, which contained overlapping partial sequences of PSTVd cDNA, were used by Cress et al. (1983) to ligate specific fragments and thereby reconstruct full-length double-stranded PSTVd cDNAs.

Preparation of cDNA probes

Radioactive and non-radioactive labels can be added to double-stranded cDNA by 'nick translation' (Rigby et al., 1977; Vivian, this volume). The method outlined in Appendix 1 (Method A) is used for preparing probes with ^{32}P-labelled nucleotides. It can also be used with some modifications to label cDNA with biotinylated nucleotides (Appendix 1, Method B).

Single-stranded cDNA probes

cDNA for a number of viroids has been prepared successfully by direct synthesis on purified viroid molecules (Palukaitis & Symons, 1979; Randles & Palukaitis, 1979; Palukaitis et al., 1981; Imperial et al., 1981). A prerequisite is several micrograms of highly purified viroid nucleic acid, free from contaminating host RNA (Symons, 1984). After incubating the viroid RNA with S1 nuclease which cleaves only a few internucleotide bonds in each molecule, a short polyadenylate tail is synthesized on the 3′-OH end of each fragment (Sippel, 1973). To form cDNA probes labelled with ^{32}P the fragments are incubated with labelled and unlabelled dNTPs in the presence of reverse transcriptase.

Single-stranded cDNA in two orientations ('plus' and 'minus') have been produced for ASBVd (Barker et al., 1985) and PSTVd (D.E. Cress, pers. comm.) using full-length double-stranded cDNA inserted into the M13mp9 phage according to the procedures described by Messing (1983). Two methods of preparing radioactive probes for ASBVd (Barker et al., 1985) and PSTVd (Salazar et al., 1988) have been compared. Method A uses a downstream primer on the M13 'plus' cDNA clone as a template to synthesize a single-stranded cDNA radioactive probe specific to the viroids. Method B uses an upstream primer on an M13 'minus' clone to produce a probe which is single-stranded in the region of the insert and double-stranded in a portion of the phage vector. The double-stranded region contains the radioactive label.

Comparison of both methods indicates that the probe prepared by method A is highly specific and less susceptible to background reactions.

RNA probes   Melton et al. (1984) constructed RNA probes by transcribing plasmid DNA templates containing a promoter for bacteriophage SP6 polymerase. The insertion of the required cDNA sequence into vectors flanked by both SP6 and T7 polymerase promoters allows the construction of probes specific for 'plus' or 'minus' sequences. Salazar et al. (1988) inserted a full-length double-stranded PSTVd cDNA into plasmid pSP65, thereby creating a template for synthesis of RNA probes for PSTVd. A method for ^{32}P-labelling of RNA transcripts is given in Appendix 2.

Comparison of   Four radioactive-labelled probes were compared for their sensi-
probes   tivity of detection of PSTVd. RNA probes were found to be the most sensitive, detecting as little as 0.33 pg of PSTVd, followed by probes prepared by Method A (M13 primer extension plus-sense insert), and nick translation of double-stranded cDNA. Probes prepared by Method B (M13 primer extension minus-sense insert) were the least sensitive (Salazar et al., 1988). RNA probes are currently used at CIP for both viroid and virus detection.

Detection of viruses and viroids with cDNA probes

Sample preparation
Viroids: The protocols of Owens & Diener (1981) and Salazar *et al.* (1988) are adequate for the detection of PSTVd and other viroids where these are found in high concentration in plant tissue. The tissue is triturated in a ratio of 1 g tissue to 2 ml of a 1:1 (v/v) mixture of formaldehyde (37%) and 10× conc. SSC (SSC is 0.15 M NaCl, 0.015 M sodium citrate, pH 7.0). The homogenate is then mixed with an equal volume of chloroform and water-saturated phenol (1:1, v/v) before centrifugation, or standing overnight at 4°C, to separate the aqueous phase.

It is not necessary to deproteinize (Sambrook *et al.*, 1989) leaf samples by addition of chloroform and phenol to detect PSTVd with radioactive probes. However, deproteinization is necessary when testing botanical or true potato seed (TPS) and potato tuber flesh or sprouts. Also, leaf samples must be thoroughly deproteinized if biotin-labelled probes are to be used, otherwise non-specific reactions may occur.

At CIP, leaf or sprout samples are routinely collected and crushed with buffer in plastic bags. Samples are transferred to test tubes, mixed with an equal volume of phenol:chloroform (1:1, v/v) and left until the aqueous phase separates. The mixing can also be done in the bags if they are resistant to solvents.

Samples of TPS are soaked for at least 2 h (usually overnight) in distilled water before crushing in a mortar as for leaf or sprout samples. A single sample can contain from one to 100 seeds, depending on the size of the stock to be tested (Salazar *et al.* 1988).

To detect viroids which occur in low concentration in plant tissue such as ASBVd, nucleic acid must be extracted carefully from large amounts of tissue. The best method, which was described by Palukaitis & Symons (1980) and modified subsequently by Allen & Dale (1981) and Barker *et al.* (1985), is described in detail in Appendix 3.

Viruses: Boulton *et al.* (1986) successfully detected virus RNA in extracts of undiluted plant sap, and at CIP 0.1−0.5 g leaf samples extracted in two volumes of 2× conc. SSC have given consistent results (L. Salazar & M. Querci, unpublished results). By simply touching a recently cut section of stem or rolled leaves onto a membrane enough sap was obtained to detect sweet potato feathery mottle virus and some other potato viruses (J. Abad & J. Moyer, pers. comm.).

Spotting of samples
Nitrocellulose membranes are rinsed thoroughly in distilled water, avoiding the formation of air bubbles, and then washed immediately with two or three changes of 5× −20× conc. SSC.

After drying on filter paper at room temperature they are stored until required in a dessicator over silica gel (note that the membranes must never be handled with bare hands).

Each spot on the membrane, located by prestamped vertical and horizontal numbers, receives 3−5 µl of the sample supernatant, usually delivered by micropipettes fitted with disposable tips, although other devices such as Pasteur pipettes or capillary tubes can be used. Once spotted the membranes are usually baked for 1 h at 80°C but good results have been obtained with only a few minutes baking (L. Salazar & M. Querci, unpublished results).

Membranes can be spotted with samples and stored either until more samples are processed and added or until the probe is ready, thereby making the best use of the ^{32}P-labelled probes which have a relatively short half-life. Membranes spotted with samples can also be sent through the post, a procedure employed at CIP to support testing for viroids or viruses by NASH in developing countries. The same procedure is also used for samples to be tested by nitrocellulose membrane-ELISA (Lizarraga & Fernandez-Northcote, 1989) or non-radioactive NASH.

Hybridization   Full details of the hybridization procedures used for the ^{32}P-labelled DNA probes, biotinylated DNA probes and RNA probes are given in Appendix 4, Methods A−C. The longest procedure takes less than a week and the others 1−2 days.

Conclusions: future prospects

Virus detection technology has changed greatly in the last 10 years with the development of ELISA and the use of recombinant DNA. Manipulation of hybridization conditions has improved the detection, and the sensitivity of detection, of viroids and viruses. For example, selection of appropriate stringency conditions during hybridization can help to control the specificity of the probe to a large extent. NASH and ELISA are generally complementary with similar sensitivities. NASH can be used in situations where ELISA cannot, but it has the disadvantage of employing radioactive labels. This shortens the useful life of the probes and restricts their use to countries where radioactive labels can be obtained and handled. Non-radioactive labels are being increasingly used but non-specific reactions with sap extracts can interfere with the results. Samples therefore require further purification which increases the labour and time involved. New labels will undoubtedly be developed and new procedures for more expeditious handling of samples are being investigated in several laboratories around the world. Furthermore, with NASH, unlike ELISA, it is possible to select and prepare probes which detect specific parts of the virus genome.

References

Allen R.N. & Dale J.L. (1981) Application of rapid biochemical methods for detecting avocado sunblotch disease. *Annals of Applied Biology* **98**, 451–61.

Barker J.M., McInnes J.L., Murphy P.J. & Symons R.H. (1985) Dot-blot procedure with (^{32}P) DNA probes for the sensitive detection of avocado sunblotch and other viroids in plants. *Journal of Virological Methods* **10**, 87–98.

Baulcombe D.C., Boulton R.E., Flavell R.B. & Jellis G.J. (1984 a) Recombinant DNA probes for detection of viruses in plants. *Proceedings of the 1984 British Crop Protection Conference*, pp. 207–13. British Crop Protection Council, Thornton Heath, Surrey.

Baulcombe D.C., Flavell R.B., Boulton R.E. & Jellis G.J. (1984 b) The sensitivity and specificity of a rapid nucleic acid hybridisation method for the detection of potato virus X in crude sap samples. *Plant Pathology* **33**, 361–70.

Boulton R.E., Jellis G.J., Baulcombe D.C. & Squire A.M. (1984) The practical application of complementary DNA probes to virus detection in a potato breeding programme. *Proceedings of the 1984 British Crop Protection Conference*, pp. 177–80. British Crop Protection Council, Thornton Heath, Surrey.

Boulton R.E., Jellis G.J., Baulcombe D.C. & Squire A.M. (1986) The application of complementary DNA probes to routine virus detection with particular reference to potato viruses. In Jones R.A.C. & Torrance L. (eds) *Developments and Applications in Virus Testing*, pp. 41–53. Association of Applied Biologists, Wellesbourne.

Clark M.F. & Adams A.N. (1977) Characteristics of the microplate method of enzyme-linked immunosorbent assay for the detection of plant viruses. *Journal of General Virology* **34**, 475–83.

Cress D.E., Kiefer M.C. & Owens R.A. (1983) Construction of infectious potato spindle tuber viroid cDNA clones. *Nucleic Acids Research* **11**, 6821–35.

Diener T.O. (1979) Viroids: structure and function. *Science* **205**, 859–66.

Gallitelli D., Hull R. & Koenig R. (1985) Relationships among viruses in the tombusvirus group: nucleic acid hybridisation studies. *Journal of General Virology* **66**, 1523–31.

Harris P.S., James C.M., Liddell A.D. & Okeley E. (1984) Viroid detection in potato quarantine: the cDNA probe and other methods. *Proceedings of the 1984 British Crop Protection Conference*, pp. 187–91. British Crop Protection Council, Thornton Heath, Surrey.

Harrison B.D. & Robinson D.J. (1982) Genome reconstitutions and nucleic acid hybridization as methods of identifying particle-deficient isolates of tobacco rattle virus in potato plants with stem-mottle disease. *Journal of Virological Methods* **5**, 255–66.

Hopp H.E., Giavedoni L., Mandel M.A. *et al.* (1988) Biotinylated nucleic acid hybridization probes for potato virus detection. *Archives of Virology* **103**, 231–41.

Hull R. (1986) The potential for using dot-blot hybridisation in the detection of plant viruses. In Jones R.A.C. & Torrance L. (eds) *Developments and Applications in Virus Testing*, pp. 3–12. Association of Applied Biologists, Wellesbourne.

Imperial J.S., Rodriguez J.B. & Randles J.W. (1981) Variation in the viroid-like RNA associated with cadang-cadang disease: evidence for an increase in molecular weight with disease progress. *Journal of General Virology* **56**, 77–85.

Leary J.L., Brigati D.J. & Ward D.C. (1983) Rapid and sensitive colorimetric method for visualizing biotin-labeled DNA probes hybridized to DNA or RNA immobilized on nitrocellulose: Bio-blots. *Proceedings of the National Academy of Science* **80**, 4045–9.

Linthorst H.J.M. & Bol J.F. (1986) cDNA hybridisation as a means of detection of tobacco rattle virus in potato and tulip. In Jones J.A.C. & Torrance L. (eds) *Developments and Applications in Virus Testing*, pp. 25–39. Association of Applied Biologists, Wellesbourne.

Lizarraga C. & Fernandez-Northcote E.N. (1989) Detection of potato viruses X and Y in sap extracts by a modified indirect enzyme-linked immunosorbent assay on nitrocellulose membranes (NCM-ELISA). *Plant Disease* **73**, 11–14.

Maule A.J., Hull R. & Donson J. (1983) The application of spot hybridization to the detection of DNA and RNA viruses in plant tissues. *Journal of Virological Methods* **6**, 215–24.

Melton D.A., Krieg P.A., Rebagliati M.R., Maniatis T., Zinn K. & Green M.R. (1984) Efficient *in vitro* synthesis of biologically active RNA and RNA hybridization probes from plasmids containing a bacteriophage SP6 promoter. *Nucleic Acid Research* **12**, 7035–56.

Messing J. (1983) New M13 vectors for cloning. *Methods in Enzymology* **101**, 20–78.

Owens R.A. & Cress D.E. (1980) Molecular cloning and characterization of potato spindle tuber viroid cDNA sequences. *Proceedings of the National Academy of Science* **77**, 5302–6.

Owens R.A. & Diener T.O. (1981) Sensitive and rapid diagnosis of potato spindle tuber viroid disease by nucleic acid hybridisation *Science* **213**, 670–2.

Palukaitis P. & Symons R.H. (1979) Hybridization analysis of chrysanthemum stunt viroid with complementary DNA and the quantitation of viroid RNA sequences in extracts of infected plants. *Virology* **98**, 238–45.

Palukaitis P. & Symons R.H. (1980) Purification and characterization of the circular and linear forms of chrysanthemum stunt viroid. *Journal of General Virology* **46**, 477–89.

Palukaitis P., Rakowski A.G., Alexander D.M. & Symons R.H. (1981) Rapid indexing of the sunblotch disease of avocados using a complementary DNA probe to avocado sunblotch viroid. *Annals of Applied Biology* **98**, 439–49.

Randles J.W. & Palukaitis P. (1979) *In vitro* synthesis and characterization of DNA complementary to cadang-cadang-associated RNA. *Journal of General Virology* **43**, 649–62.

Rigby P.W.J., Dieckmann M., Rhodes C. & Berg P. (1977) Labelling deoxyribonucleic acid to high specific activity *in vitro* by nick translation with DNA polymerase I. *Journal of Molecular Biology* **113**, 237–51.

Rosner A. & Bar-Joseph M. (1984) Diversity of citrus tristeza virus strains indicated by hybridization with cloned cDNA sequences. *Virology* **139**, 189–93.

Salazar L.F., Owens R.A., Smith D.R. & Diener T.O. (1983) Detection of potato spindle tuber viroid by nucleic acid spot hybridization: evaluation with tuber sprouts and true potato seed. *American Potato Journal* **60**, 587–97.

Salazar L.F., Balbo I. & Owens R.A. (1988) Comparison of four radioactive probes for the diagnosis of potato spindle tuber viroid by nucleic acid spot hybridization. *Potato Research* **31**, 431–42.

Sambrook J., Fritsch E.F. & Maniatis T. (1989) *Molecular Cloning: A Laboratory Manual*, 2nd edn. Cold Spring Harbor Laboratory, Cold Spring Harbor, New York.

Singh R.P. & Boucher A. (1987) Electrophoretic separation of a severe from mild strains of potato spindle tuber viroid. *Phytopathology* **77**, 1588–91.

Sippel A.E. (1973) Purification and characterization of adenosine triphosphate: ribonucleic acid adenyl transferase from *Escherichia coli*. *European Journal of Biochemistry* **37**, 31–40.

Spiegelman S. (1964) Hybrid nucleic acids. *Scientific American* **228**, 34–42.
Symons R.H. (1984) Diagnostic approaches for the rapid and specific detection of plant viruses and viroids. In Kosuge T. & Nester E.W. (eds) *Plant–Microbe Interactions: Molecular and Genetic Perspectives*, Vol. 1, pp. 93–124. Macmillan, New York.

Appendix 1: Labelling of double-stranded cDNA probes

The laboratory manual *Molecular Cloning* by Sambrook *et al.* (1989) is an invaluable source of information about the reagents and techniques described here.

METHOD A: Radioactive labelling of cDNA probes by nick-translation

Materials

- 0.2 mM deoxyribonucleoside triphosphates (dNTPs): deoxyadenosine triphosphate (dATP), deoxycytidine triphosphate (dCTP), deoxyguanosine triphosphate (dGTP) and deoxythymidine triphosphate (dTTP).
- Nick-translation buffer (10× conc.)
    0.5 M Tris-HCl, pH 7.8
    50 mM MgCl$_2$
    100 mM 2-mercaptoethanol
    100 µg ml^{-1} nuclease-free bovine serum albumin (BSA).
- DNA polymerase I/DNase I mixture
    0.4 µg µl^{-1} DNA polymerase I/40 pg µl^{-1} DNase I in 50 mM Tris-HCl, pH 7.5
    5 mM magnesium acetate
    1 mM 2-mercaptoethanol
    50% (v/v) glycerol
    100 µg ml^{-1} BSA.
- Substrate DNA: 1 µg at a concentration of 0.1–0.5 µg µl^{-1}.
- Stop buffer: 300 mM EDTA, pH 8.0.
- Sterile distilled water (SDW).
- Labelled dNTP: e.g. α [^{32}P]dCTP, ~24 TBq mmol^{-1} in aqueous solution (370 MBq ml^{-1}).
- Yeast transfer RNA: 20 mg ml^{-1} stock solution.
- 20% SDS.
- 7.5 M ammonium acetate.
- 3 M sodium acetate, pH 5.2.
- Buffered phenol (0.1 M Tris-HCl, pH 7.6).
- Chloroform.
- T$_{10}$E$_1$ buffer: 10 mM Tris-HCl, pH 7.5, containing 1 mM EDTA.

Method

- Pipette the following into a 1.5 ml Eppendorf tube sitting on ice:
    5 µl 10× conc. nick-translation buffer
    5 µl each of 0.2 mM dGTP, dATP, dTTP (or all nucleotides except the labelled one)
    1 µg substrate DNA
    156 pmol labelled dNTP (2.4 MBq).
- Make up to 45 µl with SDW.
- Mix and add 5 µl of the DNA polymerase I/DNase I mixture.
- Mix again, centrifuge and then incubate at 16°C for 1 h.
- Add, in order:
    3 µl H$_2$O

5 µl stop buffer
1 µl 20% SDS
1 µl yeast transfer RNA
30 µl 7.5 M ammonium acetate.

- Extract with an equal volume of phenol:chloroform (1:1, v/v), and transfer the aqueous phase into a clean Eppendorf tube.
- Add 225 µl ethanol and precipitate at −70°C for 1 h or at −20°C overnight.
- Centrifuge for 15 min, remove the supernatant with a pipette and resuspend the pellet in 90 µl $T_{10}E_1$. Add 10 µl of 3 M sodium acetate and 225 µl ethanol and precipitate at −70°C for 1 h or at −20°C overnight.
- Recover the pellet by centrifugation, dry it under vacuum and resuspend in 100 µl $T_{10}E_1$.

Measure the incorporation of label into the probe by trichloroacetic acid (TCA) precipitation as follows.

Materials
- 5% TCA in 0.02 M sodium pyrophosphate.
- 70% ethanol.
- 95% ethanol.
- Whatman 3 MM filter paper cut into 2 cm × 0.5 cm pieces.
- Scintillation counter.

Method
- Spot 1 µl of the sample to be assayed onto the centre of a 2 cm × 0.5 cm piece of Whatman 3 MM filter paper and let it dry completely.
- Wash the paper strip three times in ice cold 5% TCA, once in 70% ethanol and once in 95% ethanol (5 min per wash).
- Dry under a lamp and put the paper strip into a scintillation vial containing toluene-based scintillation fluid. Measure the radioactivity in a liquid scintillation counter and calculate the µl of probe to be used according to the required c.p.m. ml^{-1}.

METHOD B: Biotin labelling of cDNA probes by nick-translation

The procedure is the same as for radioactive labelling (Method A) except for the following.
- 2.5 µl of 0.4 mM labelled nucleotide, either biotin-7 dATP (Gibco-BRL) or another biotinylated nucleotide, is used in a 50 µl reaction mixture with 1 µg of substrate DNA.
- The mixture is incubated at 15°C for 90 min.
- The biotinylated DNA *must not* be extracted with phenol: chloroform. Separate it from the unincorporated nucleotides by two ethanol precipitations in the presence of 2.5 M ammonium acetate. Resuspend it in 100 µl of SSC (0.15 M NaCl, 0.015 M sodium citrate, pH 7.0). The labelled probe can be stored at 4°C for a few weeks or at −20°C for several months.

Appendix 2: Radioactive labelling of RNA probes

Materials
- 5× conc. transcription buffer
  200 mM Tris-HCl, pH 7.5
  30 mM $MgCl_2$

10 mM spermidine
50 mM NaCl.
- 10 mM stock solutions of rNTPs: ATP, CTP, UTP, GTP, pH 7.0.
- 100 mM dithiothreitol (DTT).
- Ribonuclease inhibitor, 25 units μl^{-1} (e.g. RNasin from PROMEGA).
- Labelled NTP, e.g. α [^{32}P]UTP in aqueous solution (370 MBq ml^{-1}; 24 TBq mmol^{-1}).
- SDW treated with 0.1% diethylpyrocarbonate (DEPC) before autoclaving (Sambrook et al., 1989).
- SP6 RNA polymerase or T7 RNA polymerase, 20 units μl^{-1} (PROMEGA). (Use appropriate polymerase according to RNA polymerase promoter contained in the plasmid used.)
- Template plasmid with insert, previously linearized with the appropriate restriction enzyme downstream from the insert and resuspended in $T_{10}E_1$, at a final concentration of 0.5 μg μl^{-1}.
- DNase I, 1 unit μl^{-1} (PROMEGA).

Method
- Pipette the following into a sterile centrifuge tube at room temperature.
  > 4 μl 5× conc. transcription buffer.
  > 2 μl of 100 mM DTT.
  > 0.8 μl ribonuclease inhibitor (final concentration 1 unit μl^{-1}).
  > 1 μl each of 10 mM ATP, CTP and GTP solutions.
  > 0.3 μl of 10 mM UTP.
  > 2 μl linearized plasmid template DNA (1 μg).
  > 5 μl α [^{32}P]UTP (1.85 MBq).
  > 1 μl RNA polymerase.

Add DEPC-treated SDW to give a final volume of 20 μl.
- Mix carefully, centrifuge briefly and incubate at 38°C for 90 min.
- Add 1 μl of DNase I (final concentration 1 unit μg^{-1} DNA) and incubate at 37°C for 20 min. Add 179 μl of DEPC-treated SDW.
- Extract with an equal volume of phenol:chloroform (1:1, v/v). Recover the aqueous phase and extract again with an equal volume of chloroform. Recover the aqueous phase again and precipitate RNA by adding 1/10 volume 3 M sodium acetate, pH 5.2, and 2.5 volumes of cold ethanol. Incubate at −70°C for at least 30 min.
- Centrifuge for 15 min in a microcentrifuge and resuspend the pellet in 50 μl of $T_{10}E_1$ containing 1% 2-mercaptoethanol. Store at −70°C until required.

Before hybridization, measure the incorporation of the labelled nucleotide as described in Appendix 1 Method A and calculate the amount of probe to be used.

Appendix 3: Extraction of viroids present at low concentration in host tissue

Materials
- TSS buffer
  > 0.1 M Tris-HCl, pH 8.5
  > 0.5 M NaCl
  > 0.5% SDS

2.5% PVP-40

0.1 mM magnesium acetate

Add 1% DIECA immediately before use.

- Tris-HCl buffer

    75 mM Tris-HCl, pH 7.0

    0.15 M NaCl

    1.5 mM EDTA.

- Sodium acetate buffer

    0.5 M sodium acetate, pH 5.0

    0.2 M NaCl

    1 mM EDTA.

- Cold ethanol.

- Chloroform:phenol (1:1, v/v).

- 12 mM LiCl.

- 1.0 M magnesium acetate.

Methods
- 25–30 g of tissue are triturated for 1 min at room temperature in 90 ml of TSS buffer.
- NaCl (7.9 g) is added and the mixture is triturated for a further minute.
- Incubate at −15°C for 1 h.
- Centrifuge at 10 000 g for 20 min at 2°C.
- Add 2 volumes of cold ethanol to 50 ml of supernatant.
- Incubate at −15°C for 1 h.
- Centrifuge to recover the nucleic acids and resuspend the pellet in 12.5 ml of Tris-HCl buffer.
- Add an equal volume of chloroform:phenol and centrifuge to separate the aqueous phase.
- Nucleic acids are precipitated from the aqueous phase with an equal volume of 12 mM LiCl and 0.01 volume of 1.0 M magnesium acetate.
- Centrifuge at 10 000 g for 20 min.
- Resuspend the pellet in 10 ml sodium acetate buffer.
- Precipitate the nucleic acids once more by adding 2.5 volumes of cold ethanol.
- Centrifuge, dry and resuspend the pellets in 0.5 ml of 0.1 mM EDTA.
- Store at −20°C.

Appendix 4: Hybridization protocols

METHOD A:
Hybridization with
[³²P]DNA probes
Materials
- Hybridization solution

    50% deionized formamide

    5× conc. SSC

    50 mM NaPO4, pH 6.5

    2.5 mM EDTA

    0.6% SDS

    5× conc. Denhardt's solution.

- 50% Dextran sulphate.
- ³²P-labelled DNA probe (specific activity should be about 108 c.p.m. μg^{-1}).
- 10 mg ml^{-1} Herring Sperm DNA.

- SSC/SDS wash buffers
    I: 2× conc. SSC/0.1% SDS
    II: 0.2× conc. SSC/0.2% SDS
    III: 0.1× conc. SSC/0.2% SDS.

Method
- Put the membrane in a suitable heat-sealable polythene bag.
- For a 12 cm × 16 cm membrane add 9 ml of hybridization solution (adjust the volume according to the membrane size — a suitable ratio is at least 1 ml of solution per 20 cm^2).
- Denature the herring sperm DNA by heating it at 100°C for 10 min. Chill on ice and add 120 μl to the bag (final conc. 120 μg ml^{-1})
- Add 1 ml of 50% dextran sulphate (final concentration 5%) and mix. Incubate for 2 h at 55°C for viroids (45°C for viruses).
- Denature the correct amount of probe (about 1–2.5 × 10^6 c.p.m. ml^{-1}) by heating at 100°C for 7 min. Chill quickly on ice and add to the hybridization bag. Seal and avoid trapping air bubbles.
- Hybridize for 18–24 h at 55°C for viroids and 45°C for viruses.
- Recover the membrane from the bag. Dispose of the buffer safely and remember that it is highly radioactive.
- Wash the membrane in a tray on a rotary shaker:
    twice for 15 min at room temperature in SSC/SDS I;
    twice for 15 min at 37°C in SSC/SDS II;
    twice for 15 min at 55°C in SSC/SDS III;
    final rinse in 0.1 × conc. SSC.
- Allow the membrane to dry on tissue paper at room temperature and then autoradiograph for 24–48 h at −70°C with Kodak X-Omat film (or similar) using an intensifying screen such as Dupont Cronex Lightning Plus.

METHOD B:
Hybridization with
biotinylated
DNA probes
Materials
- Prehybridization solution
    50% deionized formamide
    5× conc. SSC
    5× conc. Denhardt's solution (0.1% Ficoll, 0.1% PVP, 0.1% BSA)
    25 mM sodium phosphate, pH 6.5
    0.5 mg ml^{-1} freshly denatured, sheared herring sperm DNA.
- Hybridization solution
    45% formamide
    5× conc. SSC
    1× conc. Denhardt's solution
    20 mM sodium phosphate, pH 6.5
    5% dextran sulphate
    0.2 mg ml^{-1} freshly denatured, sheared herring sperm DNA
    0.1–0.5 μg ml^{-1} freshly denatured, biotinylated DNA probe.
- Buffer 1
    0.1 M Tris-HCl, pH 7.5
    0.15 M NaCl.
- Buffer 2
    0.1 M Tris-HCl, pH 9.5
    0.1 M NaCl
    0.05 M MgCl$_2$.

- Streptavidin–alkaline phosphatase conjugate (SA–AP).
- Nitro blue tetrazolium chloride, grade III (NBT).
- 5-bromo-4-chloro-3-indolyl phosphate (BCIP).
- Blocking solution: 3% (w/v) BSA (fraction V) in buffer 1.
- SSC/SDS wash buffers:
    I; 2× conc. SSC/0.1% SDS
    II; 0.2× conc SSC/0.1% SDS
    III; 0.16× conc. SSC/0.1% SDS.

Method
- The prehybridization and hybridization stages are again performed in heat-sealable polythene bags with approximately 1 ml of solution per 20 cm^2 of membrane.
- Denature sheared herring sperm DNA by heating at 100°C for 10 min, chill immediately on ice, add to the prehybridization solution and mix well.
- Pour the appropriate amount of prehybridization solution into the bag, seal without trapping air bubbles and incubate in a water bath at 42°C for 2 h.
- Replace the prehybridization solution with the same amount of hybridization solution containing denatured herring sperm DNA and an adequate amount of biotinylated probe, which is denatured and chilled on ice just before use.
- Seal the bag, mix well and allow to hybridize at 42–45°C (the hybridization solution containing the probe can be reused if stored at −20°C and denatured again before use).
- The membrane is placed in a plastic tray and washed with 250 ml of solution at each step on a rotary shaker. Washing is carried out at room temperature (if not otherwise indicated) as follows:
    twice for 5 min in SSC/SDS I;
    twice for 5 min in SSC/SDS II;
    twice for 15 min at 50°C in SSC/SDS III;
    final rinse in 2× conc. SSC buffer.
- Incubate the membrane at 60°C for 1 h with 50 ml of preheated blocking solution in a covered plastic tray.
- Incubate on a rotary shaker for 20 min at room temperature with 20 ml buffer 1 containing SA–AP conjugate, diluted 1:1000.
- Wash twice for 15 min in buffer 1 containing 0.05% Tween 20, and then for 10 min in buffer 2.
- Transfer to a clean tray with a lid containing the colour development solution made up as follows: 6 mg NBT and 3 mg BCIP in 30 ml of buffer 2 (both can be kept as stock solutions in dimethyl formamide). *Wear gloves throughout this procedure.*
- Incubate on a shaker in the dark until colour develops and stop the reaction by rinsing in distilled water.

METHOD C: Hybridization with RNA probes Materials
- Hybridization solution
    40% deionized formamide
    0.18 M NaCl
    10 mM sodium cacodylate
    1 mM EDTA

0.1% SDS.
- 50% Dextran sulphate.
- Calf thymus DNA 4 mg ml^{-1}.
- Wash buffer 1
  0.36 M NaCl
  10 mM Tris-HCl, pH 7.5
  0.1% SDS.
- Wash buffer 2
  0.1× conc. SSC
  0.1% SDS.
- Wash buffer 3
  2× conc. SSC.
- RNase A stock solution, 10 mg ml^{-1}.

Method
- Place the membrane in a polythene bag and add 9.6 ml of hybridization solution (enough for a 12 cm × 16 cm membrane).
- Incubate at 55°C for 10 min.
- Denature the calf thymus DNA by heating at 100°C for 5 min, chill on ice and add 1 ml to the bag to give a final concentration of 300 μg ml^{-1}.
- Incubate again at 55°C for 10 min.
- Add 2.4 ml of 50% dextran sulphate and incubate at 55°C for 10 min.
- Add enough probe to give approximately 400 000 c.p.m. ml^{-1} of hybridization solution.
- Immerse the bag overnight in a water bath at 55°C for viroids and 45°C for viruses.
- Remove the membranes and wash as follows:
    twice for 20 min with wash buffer 1 at room temperature;
    once for 30 min with wash buffer 2 at 65°C;
    twice for 10 min with wash buffer 3 at room temperature;
    once with wash buffer 3 containing RNase A at a final concentration of 2 μg ml^{-1} at room temperature.
- Dry the membrane over tissue paper at room temperature or under an incandescent lamp.
- Autoradiograph at −70°C overnight using Kodak X-Omat AR (or similar) with an intensifying screen.

8 Identification of plant pathogenic bacteria using nucleic acid technology

A. VIVIAN

Introduction

The genetic blueprint for each bacterium, its genome, comprises all the information in code form to determine its precise characteristics, including its specificity as a plant pathogen. Therefore, at least in theory, what better idea than to use the specificity that must reside in the DNA of the organism to identify it? However, like many theoretically compelling notions, the practice is rather more problematical. This review will attempt to explain the range of methodologies that are currently available for the identification of plant pathogenic bacteria using DNA technology. It will also give some examples of the practical applications of such techniques.

We live in an age of increasingly rapid change and nowhere is this more apparent than in the field of agriculture and horticulture. Often, moves towards intensification, involving changes of scale and the abandonment of traditional practices, have led to the emergence of novel diseases or of old diseases on a greatly increased scale. The increased potential for the transmission of alien diseases on imported seed, such as the introduction of bacterial blight of pea to the UK in 1985 (Stead & Pemberton, 1987), together with the subsequent quarantine and certification of disease-free seed stocks, highlights the need for rapid and effective detection of the pathogen concerned. At present such procedures are often lengthy, involving the culture and testing of organisms on plants (Ball & Reeves, this volume). Nucleic acid technology offers the prospect of greater rapidity and sensitivity of detection and improved specificity in diagnosis.

Methodologies: a glossary of terms

Inevitably in any rapidly developing area of research there is a mass of new terminology generated, some of it not always consistent in its application. In this article I propose to use the term genomic fingerprinting for DNA band analysis that does not involve the use of hybridization with a nucleic acid probe, and to reserve the term restriction fragment length polymorphism (RFLP) analysis for DNA band analysis with a nucleic acid probe.

145

DNA: double- and single-stranded

Bacteria possess a single chromosome comprising a circular double-stranded DNA molecule, with the addition in many cases (but not all) of smaller circular double-stranded DNAs called plasmids. These DNA molecules can be exposed by lysis of the bacterial cell, and subsequently denatured under appropriate ionic conditions that result in separation of the complementary single strands of the duplexes. By manipulation of the buffer conditions and temperature, re-establishment of the pairing between complementary strands can lead to annealing and regeneration of double-stranded molecules.

Genomic fingerprinting

One of the earliest uses of the very precise specificity of DNA restriction enzyme cleavage was the so-called genomic fingerprinting of bacteria as a means of identification (Kaper et al., 1982; Bradbury et al., 1984; Kristiansen et al., 1984). Total genomic DNA is digested with one or more restriction enzymes and the resulting fragments are separated by agarose (or polyacrylamide) gel electrophoresis. The resulting DNA bands can be stained with ethidium bromide and viewed on an ultraviolet transilluminator. Band patterns can be recorded by photography for analysis and comparison (see Appendix 1).

Field inversion gel electrophoresis

Fragments of DNA resulting from digestion with restriction enzymes can be efficiently separated by simple electrophoresis, up to a size limit of about 20 kb. Linear DNA molecules larger than about 20 kb move through the gel medium at mobilities nearly independent of their size. This is thought to be associated with wedge (>)-shaped migration of the molecule in one direction. The principle behind field inversion is that if the molecules are periodically forced to reverse their direction of travel, this requires a series of conformational changes that must propagate from one end of the molecule to the other during inversion of the 'wedge' ($> \longrightarrow <$). Adjustment of the periods of current inversion by a suitable switching device permits the effective separation of molecules up to about 700 kb (Carle et al., 1986).

Colony blotting

The precision with which bacterial strains can be identified will, to some extent, depend on the time available to conduct the test. The most rapid way to obtain a sample of total genomic DNA ready for probing is to use the colony blotting technique. This involves growing the isolated bacteria on an agar plate. The resulting colonies can then be transferred, using a replica-plating procedure, to a cellulose nitrate or nylon membrane, overlaying a further agar plate. Overnight incubation is usually sufficient to obtain satisfactory growth of the colonies on the surface of the membrane. The membrane is then treated so as to lyse the bacterial cells, denature the released DNA and fix the DNA to the filter (see Appendix 2). Nylon membranes have the advantage that they can be probed more than once using different probe DNA on each occasion, although there is some loss of quality with succeeding hybridizations.

Probe DNA The process of DNA strand separation and re-annealing can be used to hybridize related (complementary) single-stranded DNA molecules to a specific DNA sequence. Such a cloned sequence is referred to as a DNA probe. Depending upon the circumstances of use the probe may comprise of a linear sequence of DNA or the complete cloning vector plus insert sequence.

Prehybridization To prevent a high level of non-specific binding of the DNA probe to the membrane, it is essential to treat the membrane with a non-specific DNA such as that from salmon sperm which has been boiled and sheared. This will avoid problems of high background (Appendix 2).

Hybridization The prepared, labelled DNA probe and membrane are then incubated together under suitable conditions to promote the complementary annealing of single-stranded DNAs. Excess DNA probe is removed by a series of washes, the rigour of which determine the stringency (or degree of specificity) of the hybridization.

Nick-translation Using a procedure called nick-translation (Rigby *et al.*, 1977) it is possible to introduce a series of radio-labelled or chemically-labelled nucleotide bases into a specific probe DNA molecule. This makes use of the DNA polymerase I enzyme of *Escherichia coli* to catalyse the addition of bases to the 3′-hydroxyl terminus of a single-stranded break (or nick) created in the duplex DNA by pancreatic DNase I. The nicks are introduced at random by this procedure, resulting in the generation of a population of partially overlapping radioactive fragments. Before use, this labelled DNA probe must be converted to a single-stranded form for hybridization (see Appendix 3).

Labelling and detection of the nucleic acid probe The most widespread method has been the enzymatic incorporation of radio-label in the form of a ^{32}P-containing deoxynucleoside triphosphate (usually cytosine) by nick-translation. The subsequent detection of the bound probe will depend upon the type of label used, whether radioactive or chemical. In the case of ^{32}P-labelled probes, detection is by autoradiography. This involves the use of X-ray film, often with intensification screens and at temperatures of −70°C to obtain a result in 24−48 h.

However, the use of radioactive sources requires dedicated and expensive facilities which may not be readily available in many routine laboratories or in field situations.

Consequently there is great interest in the development of non-radioactive chemical probes (see reviews by van Brunt & Klausner, 1987; Miller & Martin, 1988). 'Cold' probes include the use of reporter groups that can be detected directly following hybridization. They include systems based on alkaline phosphatase and peroxidase (Renz & Kurz, 1984) but these generally have poor sensitivity. An alternative is to introduce ligands into DNA that are then indirectly detected following hybridization. Biotinylated bases are incorporated by nick-translation into the probe, which is later detected with avidin—enzyme conjugates or anti-biotin antibody. There have been many variations of the use of biotin label (Walker & Dougan, 1989).

Another technique is chemical modification of bases (haptens) which can be detected after hybridization with anti-hapten antibodies (Candlish *et al.*, this volume); this includes sulphonation of cytosine residues, marketed as Chemiprobe™ (Bioproducts). A kit is used to insert antigenic sulphone groups into cytosine residues of the probe. Following hybridization, the probe is detected using a sandwich immuno-enzymatic reaction. Monoclonal antibody binds to the sulphone residues and then to an alkaline phosphatase—anti-immunoglobulin conjugate. Addition of a chromogenic alkaline phosphatase substrate produces a blue colour in the presence of the hybridized probe. Recently, Tropix has introduced a chemi-luminescent enzyme substrate for the detection of alkaline phosphatase. The compound, AMPPD (3-(2′-spiroadamantane-4-methoxy-4-(3-phosphoryloxy)-phenyl-1,2-dioxetane), is marketed in the UK by New Brunswick Scientific. Until recently cold probes have not been as sensitive nor as specific as radio-labelled probes. However, several of the recent introductions, including both AMPPD and digoxigenin (Boehringer), offer extended shelf-life of up to 1 year or more, together with a high degree of sensitivity.

Dot-blots   Another relatively quick and simple technique involves the rapid isolation of a small sample of DNA, which can then be aspirated in a simple device onto a suitable membrane for probing. Purpose-built equipment of this kind is available from Bio-Rad and many other companies (see Appendix 4). The main advantages of this method over the colony blotting procedure are the greater purity of sample that can be achieved and the ability to control the amount of DNA present in the sample.

Squash-blots   Plant material can be used directly as a source of bacteria to be detected and identified. The procedure described is that of Gilbertson *et al.* (1989) for use with bean leaves but it could be adapted to most plant tissue. Excised leaf discs, preferably from the margin(s) of suspected lesions, are squashed onto the membrane and the DNA fixed in the usual manner (see Appendix 4). An alternative procedure involves the addition of a macerated suspension of leaf discs as spots on a membrane filter. The advantage of the latter procedure is that the same suspensions could be used to determine a viable count of the bacteria present.

Restriction fragment length polymorphism   This is potentially the most unequivocal method for the identification of bacteria; it involves the isolation and enzymic digestion of total genomic DNA from a culture of the organism (Appendix 1). The separated fragments can be transferred to a membrane by Southern blotting (Sambrook *et al.*, 1989). Subsequent hybridization and detection are similar to that described for colony hybridization (Appendix 2). However, the resulting autoradiograph provides a pattern of hybridizing bands for each bacterium, which with an appropriate DNA probe can be used to differentiate organisms or to confirm the identity of two different isolates of the same organism.

Copy number of DNA sequence   Even a casual inspection of an RFLP result may immediately indicate to the observer marked differences in the strength of signal for any particular band. This may result from either an increased copy number of the sequence present in the band or it may reflect very close homology with the probe DNA sequence (or vice versa). Comparison with a photograph of the ethidium bromide-stained gel should show when the strength of hybridization correlates well with the intensity of the DNA band, indicating increased copy number as the likely explanation. As a general rule fragments which appear over-represented may result from two causes. Some sequences may be repeated (occasionally many times over) in the genome. These might, for example, include genes for ribosomal RNA (rRNA), but may also include apparently non-coding regions of DNA (Szabo & Mills, 1984). The other cause is the presence of plasmids, the largest of which may be only one to three times more frequent in copy number than the chromosome, while smaller ones might be 20−30 times more frequent.

Plasmid versus chromosomal DNA probes   Plant pathogenic bacteria frequently (but not invariably) harbour one or more plasmids which may range in size from a few kilobase pairs of DNA up to megaplasmids (>1000 kb), detected in some strains of *Pseudomonas solanacearum* (Boucher *et al.*, 1986). In general smaller plasmids tend to be present in a greater number of copies per cell than larger ones and most will exceed the number of chromosome copies per cell (see above). These observations clearly have implications for the choice of source for the probe DNA in relation to the type of analysis for which it is to be used. In the case of probes for RFLPs it is important that the probe is not too restricted in its potential to hybridize with the genomes of the organisms to be distinguished. In this instance a repeated sequence in the genome may be ideal if it results in sufficient heterogeneity in band patterns. Conversely, repeated sequences may well be found across a range of pathovars and this would make them unsuitable for use in colony- or dot-blot hybridization tests. Ideally, for pathovar distinction, one would prefer a unique sequence related intrinsically to that pathovar. For example, one could imagine that where specific toxin production is restricted to one pathovar, a probe derived from part of the structural gene sequence for the toxin might in this case be ideal. However, it is likely that such genes may be chromosomal and therefore present only in low copy number per cell. In colony- and dot-blot the sensitivity of detection is likely to be enhanced if the probe DNA is multicopy *in vivo* and therefore plasmid-borne.

Finally, the simplest kind of DNA probe to prepare and use in a routine situation will probably comprise both specific insert sequence and the cloning vector used to propagate it. Since many of these vectors are based on bacterial plasmids (usually from *E. coli*), it is important to ensure that these vector sequences do not in practice produce false-positive identifications in the test system. Separation of the cloned insert from the vector prior to use would avoid such problems but would increase the labour and complexity of using such probes. The use of designed synthetic oligonucleotide probes, although more expensive, will ultimately achieve convenience and a high degree of specificity.

Oligonucleotide synthesis

Relatively inexpensive equipment is now available for the custom synthesis of specific 20-nucleotide (20-mer) sequences of DNA (Applied Biosystems). Knowledge of the DNA sequence of the cloned region is a prerequisite to enable the selection of a specific sequence for synthesis. Such oligonucleotides have many applications including their use as primers in the polymerase chain reaction (PCR) (see below).

Polymerase chain reaction

Once a region of DNA which has potential as a DNA probe has been identified and its nucleotide sequence determined, it is possible to make use of parts of this sequence (which act as primers) to increase the concentration or amplify part or all of the sequence. This can be done by PCR, which is capable of producing millions of copies of a specific DNA sequence in a few hours (Saiki *et al.*, 1988). First, two primers are added to the denatured (single-stranded) DNA to be amplified, one specific for each end of the sequence (enabling use of both strands as templates). Following annealing of the primers to their complementary sequences, the enzyme *Taq* DNA polymerase (from *Thermus aquaticus*) is used to synthesize multiple copies of the sequence in a cyclical reaction procedure (Saiki *et al.*, 1988; Erlich, 1990; Innis *et al.*, 1990). The resultant large increase in concentration makes detection of the sequence by a specific probe much more effective. In theory, such an approach could be used in conjunction with direct washings from seed samples for the detection of certain seed-borne pathogens, by seeking out and amplifying DNA sequences that would permit identification. Unfortunately, while this approach works well with purified DNA, there are considerable problems with its use in 'dirty' samples containing large amounts of other DNA because this results in a high level of non-specific background hybridization with probe DNA.

Strategies for the identification and detection of bacteria

Genomic fingerprinting of *Xanthomonas campestris* Hartung and Civerolo (1987) used *Eco*R I digestion of total DNA to study genomic fingerprints of DNA fragments produced over a range of approximately 1–2 kb from strains of *Xanthomonas campestris* pv. *citri*. The DNA fragments were separated by polyacrylamide gel electrophoresis, stained with ethidium bromide and photographed on a transilluminator. While this approach clearly distinguished the geographically distinct Asiatic and Argentinian groups, strains from Florida were very heterogenous, even between isolates from a single disease outbreak.

In an attempt to minimize the complexity of bands usually encountered in genomic fingerprinting, Cooksey & Graham (1989) chose restriction endonucleases that cut infrequently in GC-rich genomes. The enzymes *Dra* I and *Ssp* I have recognition sequences comprising only A and T nucleotides. The relatively few large DNA fragments can be separated by field inversion gel electrophoresis. The method was used successfully with both *Pseudomonas* and *Xanthomonas* strains. Commercially available pulse controllers with programmable pulse times can be used to separate complex patterns of large DNA fragments.

The main disadvantage of genomic fingerprinting as described above is the complexity of the band patterns obtained, and the labour involved in direct comparison of photographs. The use of densitometer profiles in conjunction with computerized analysis of the results has been successfully applied to some medically important bacteria such as *Campylobacter* (Bruce *et al.*, 1988).

RFLP analysis using radio-labelled DNA probes

More recently, Hartung & Civerolo (1989) have used radio-labelled DNA probes to investigate RFLP in *Eco*R I and *Pvu* II digests of *Xanthomonas campestris* pv. *citri* total DNA. Use of these probes permitted analysis of a much more comprehensive range of DNA fragments (approx. 0.5−25 kb) and, at the same time, achieved a visually simpler pattern of bands.

Perhaps the most comprehensive attempt at using RFLP to identify plant pathogenic bacteria to date has been the work of Lazo *et al.* (1987) with *X. campestris*. Using cloned DNA fragments obtained from a strain of *X. campestris* pv. *citri*, they were able by appropriate choice of probe and restriction endonuclease to demonstrate pathovar-specific RFLP patterns for each of 26 pathovars. The two probes were library cosmids, chosen at random and containing relatively large inserts of DNA (30 and 37 kb), which were presumed to be chromosomal in origin since no plasmids were detected in the parent strain. The complexity of banding seen on the resulting autoradiographs necessitated the use of densitometric measurement.

Gabriel *et al.* (1988) subsequently investigated RFLP patterns amongst the diverse 'E' group of *X. campestris* pv. *citri* strains from the Florida outbreaks of citrus canker (Schoulties *et al.*, 1987). Their analysis indicated that one group of strains was related to *X. campestris* pv. *alfalfae*, and a subgroup of these strains had a similar host range to that pathovar. It also demonstrated the potential that RFLP analysis has for identification, without recourse to plant experiments, particularly for isolates of uncertain origin or host range.

All of the DNA probes described above have been based on DNA sequences of unknown function. Cook *et al.* (1989) developed probes based on regions of DNA known to specify virulence and hypersensitive response (*hrp* genes) for investigating the genetic diversity amongst 62 strains of *P. solanacearum*. Clearly, as we extend our knowledge of the genetic basis of pathogenicity there will be an increasing opportunity to devise probes based on a more rational approach to the types of bacteria we wish to detect.

In view of the labour and time involved in the conduct of RFLP analysis, together with the frequent complexity involved in the interpretation of the results generated, this approach is unlikely to be used in routine analysis. Its potential lies mainly in the fields of precise identification of strains in relation to epidemiology and taxonomy.

Detection and identification by colony hybridization

The essential simplicity of colony hybridization, involving only growth on an agar plate prior to analysis, offers the most rapid and reliable method at present available for routine analysis in the laboratory. Firrao (1990) used a mixture of five cloned *Sal* I fragments from a plasmid pCS1 from *Clavibacter michiganensis* subsp. *sepedonicus* as specific DNA probes for this organism (the cause of potato ring rot) and using the non-radioactive label, digoxigenin (Boehringer). *C. michiganensis* subsp. *sepedonicus* could be distinguished from *C. michiganensis* subsp. *insidiosus*, with which it shares serological determinants and virtual 100% DNA homology. The probe did not hybridize with other pathogens and saprophytes, including *Corynebacterium pyogenes*, *Arthrobacter simplex*, *X. campestris*, *P. solanacearum* or *Erwinia carotovora* subsp. *carotovora*. Loss of viability in potato samples stored at −20°C presented no problem, since viable cells were not necessary for use of the DNA probe. Johansen *et al.* (1989) have also described DNA probes for the identification of *C. michiganensis* subsp. *sepedonicus*. These were obtained from a library of randomly cloned fragments of *C. michiganensis* subsp. *sepedonicus*, one of which, a 2.6 kb *Pst* I fragment, contained a region of DNA that was repeated many times in the genome. The authors reported that while a radio-labelled DNA probe was able to detect 0.5 ng of isolated *C. michiganensis* subsp. *sepedonicus* DNA, a biotinylated probe was less sensitive and less specific because it was adversely affected by material present in potato tuber extracts. These kinds of problem necessitate the culture of bacteria prior to detection by biotinylated DNA probes.

Xanthomonas campestris pv. *phaseoli* and *X. campestris* pv. *phaseoli* var. *fuscans* cause common bacterial blight of bean (*Phaseolus vulgaris* L.). Gilbertson *et al.* (1989) investigated the potential of total plasmid DNA and two cloned fragments of plasmid DNA from *X. campestris* pv. *phaseoli* to act as DNA probes to detect both bean pathogens and to differentiate them from non-pathogenic xanthomonads associated with bean debris. Using colony hybridization and total plasmid DNA as a probe they were satisfied that the pathogenic xanthomonads could be successfully distinguished from non-pathogenic ones. However, even cloned 3.4 kb and 1.6 kb *EcoR* I fragments of plasmid DNA showed cross-reaction with DNA bands of other *X. campestris* pathovars when total DNA from these bacteria was Southern blotted and probed.

Two groups have attempted to obtain specific DNA probes for

the identification of *Pseudomonas syringae* pv. *pisi* by Southern blotting and colony hybridization (Rasmussen & Wulff, 1990; A. Vivian, H. Dewhurst, A.D. Bavage, J. Reeves & S. Ball, unpublished results). In each case, considerable cross-hybridization with other *P. syringae* pathovars was encountered.

In colony blotting there are also indications of better specificity being achieved with more rational approaches to probe design based on regions of DNA specifying known functions. Thus digoxigenin-labelling of specific DNA fragments from the gene cluster encoding production of and resistance to phaseolotoxin enabled the production of a highly specific probe for *P. syringae* pv. *phaseolicola* (C. Manceau & C. Tourte, pers. comm.). Using a similar approach Schaad *et al.* (1989) reported the use of a radio-labelled DNA probe carrying a gene involved in phaseolotoxin production which was pathovar-specific for individual colonies of *Pseudomonas phaseolicola* in pure and mixed cultures, in seed washings and in diseased specimens collected in the field.

Cuppels *et al.* (1990) have recently described the use of the Chemiprobe™ non-radioactive reporter system in conjunction with a fragment of chromosome from the region controlling production of the toxin coronatine from *P. syringae* pv. *tomato*. The detection limit was 4×10^3 CFU in squash-blots from extracted leaf material. The probe was highly specific for *P. syringae* pv. *tomato* in field screening.

DeParasis and Roth (1990) partially sequenced 16S rRNA cDNAs from 52 strains of phytobacteria including *Xanthomonas*, *Erwinia* and *Pseudomonas*. A variable region of about 40 nucleotides did not permit differentiation of pathovars or subspecies but the results indicated that the degree of variation in the conserved 16S rRNA is sufficiently high to expect specific discrimination between genera using a synthetic oligonucleotide probe.

Direct detection in plant material — use of PCR

P.R Mills (pers. comm.) isolated an 800 bp DNA fragment from a plasmid found in *Erwinia amylovora*. This probe did not cross-react with *E. herbicola* or any other epiphyte tested. Oligonucleotides corresponding to the terminal sequences of the 800 bp fragment were constructed for use as primers to permit amplification of this sequence by PCR. While it was not possible to detect the pathogen by PCR in direct plant washings, PCR was very effective in improving the sensitivity of detection with phenol-extracted samples.

Ward & De Boer (1990) described a DNA probe (approx. 1 kb) isolated from a mixture of genomic DNAs of *E. carotovora* subsp. *atroseptica* and *E. carotovora* subsp. *carotovora*, enriched for sequences lacking hybridization potential with *E. coli*. This DNA probe was used successfully in dot-blots to identify both subspecies of *E. carotovora*. The results compared favourably with those obtained from soil platings. The limit of detection in dot-blots with ^{32}P-labelled DNA probe was 20 CFU of *E. carotovora* subsp. *atroseptica* in 100 µl of soil extract containing 2×10^4 CFU.

Conclusions and future perspectives

The specificity inherent in DNA sequences provides a basis for the very precise identification of bacterial strains. The development of nucleic acid probes for particular tasks requires that the procedures be simple, rapid, unambiguous and capable of use under a wide range of conditions. To these ends, the future would appear to lie with amplification techniques such as PCR which, if the problems associated with direct testing of plant material can be overcome, would seem to offer the prospect of rapid detection and identification of pathogens without recourse to their culture. Improved chemical labelling of DNA offers the hope of a sensitive, cost-effective alternative to the use of radioisotopes.

At another level the increasing sophistication of RFLP analysis can serve many useful functions. For example, Cook *et al.* (1989) found that three distinct RFLP groups within a single race of *P. solanacearum* were each associated with different epidemics of unique geographic origin. In this way RFLP may shed light on the evolution of races and subspecies implicated in new disease outbreaks.

RFLP also appears to offer an objective means of classification for at least some plant pathogen species. The development of the technique offers the prospect of a classification for bacteria that reflects the natural evolution of species (Cook *et al.*, 1989).

References

Boucher C.A., Martinel A., Barberis P., Allgoing G. & Zischek C. (1986) Virulence genes are carried by a megaplasmid of the plant pathogen *Pseudomonas solanacearum*. *Molecular and General Genetics* **205**, 270–5.

Bradbury W.C., Pearson A.D., Marko M.A. Congi R.V. & Penner J.L. (1984) Investigation of a *Campylobacter jejuni* outbreak by serotyping and chromosomal restriction endonuclease analysis. *Journal of Clinical Microbiology* **19**, 342–6.

Bruce D., Hookey J.V. & Waitkins S.A. (1988) Numerical classification of campylobacters by DNA-restriction endonuclease analysis. *Zentralblatt für Bakteriologie, Mikrobiologie und Hygiene* **A269**, 284–97.

Carle G.F., Frank M. & Olson M.V. (1986) Electrophoretic separations of large DNA molecules by periodic inversion of the electric field. *Science* **232**, 65–8.

Cook D., Barlow E. & Sequeira L. (1989) Genetic diversity of *Pseudomonas solanacearum*: detection of restriction fragment length polymorphisms with

DNA probes that specify virulence and the hypersensitive response. *Molecular Plant—Microbe Interactions* **2**, 113–21.

Cooksey D.A. & Graham J.H. (1989) Genomic fingerprinting of two pathovars of phytopathogenic bacteria by rare-cutting restriction enzymes and field inversion gel electrophoresis. *Phytopathology* **79**, 745–50.

Cuppels D.A., Moore R.A. & Morris V.L. (1990) Construction and use of a nonradioactive DNA hybridisation probe for detection of *Pseudomonas syringae* pv. *tomato* on tomato plants. *Applied and Environmental Microbiology* **56**, 1743–9.

DeParasis J. & Roth D.A. (1990) Nucleic acid probes for identification of phytobacteria: identification of genus-specific 16S rRNA sequences. *Phytopathology* **80**, 618–21.

Erlich H.A. (1990) *PCR Technology. Principles and Applications for DNA Amplification.* Macmillan, London.

Firrao G. (1990) Cloned diagnostic probe for *Clavibacter michiganensis* ssp. *sepedonicus. EPPO Bulletin* **20**, 207–13.

Gabriel D.W., Hunter J.E., Kingsley M.T., Miller J.W. & Lazo G.R. (1988) Clonal population structure of *Xanthomonas campestris* and genetic diversity among citrus canker strains. *Molecular Plant—Microbe Interactions* **1**, 59–65.

Gilbertson R.L., Maxwell D.P., Hagedorn D.J. & Leong S.A. (1989) Development and application of a plasmid DNA probe for detection of bacteria causing common bacterial blight of bean. *Phytopathology* **79**, 518–25.

Grunstein M. & Hogness D.S. (1975) Colony hybridization: a method for the isolation of cloned DNAs that contain a specific gene. *Proceedings of the National Academy of Sciences USA* **72**, 3961–5.

Hartung J.S. & Civerolo E.L. (1987) Genomic fingerprints of *Xanthomonas campestris* pv. *citrus* strains from Asia, South America and Florida. *Phytopathology* **77**, 282–5.

Hartung J.S. & Civerolo E.L. (1989) Restriction fragment length polymorphisms distinguish *Xanthomonas campestris* strains isolated from Florida citrus nurseries from *X. c.* pv. *citri. Phytopathology* **79**, 793–9.

Innis M.A., Gelfand D.H., Sninsky J.J., & White T.J. (1990) *PCR Protocols: A Guide to Methods and Applications.* Academic Press, London.

Johansen I.E., Rasmussen O.F. & Heide M. (1989) Specific identification of *Clavibacter michiganensis* subsp. *sepedonicum* by DNA-hybridization probes. *Phytopathology* **79**, 1019–23.

Kaper J.B., Bradford H.B., Roberts N.C. & Falkow S. (1982) Molecular epidemiology of *Vibrio cholerae* in the U.S. Gulf coast. *Journal of Clinical Microbiology* **16**, 129–34.

Kristiansen B.-E., Bjorvatn B., Lund V., Lindqvist B. & Holten E. (1984) Differentiation of B15 strains of *Neisseria meningitidis* by restriction endonuclease fingerprinting. *Journal of Infectious Disease* **150**, 672–7.

Lazo G.R., Roffey R. & Gabriel D.W. (1987) Pathovars of *Xanthomonas campestris* are distinguishable by restriction fragment-length polymorphism. *International Journal of Systematic Bacteriology* **37**, 214–21.

Miller S.A. & Martin R.R (1988) Molecular diagnosis of plant disease. *Annual Review of Phytopathology* **26**, 409–32.

Rasmussen O.F. & Wulff B.S. (1989) Identification and use of DNA probes for plant pathogenic bacteria. In Christiansen C., Munck L. & Villadsen J. (eds) *Proceedings of the 5th European Congress on Biotechnology*, Vol. 2, pp. 693–8. Munksgaard, Copenhagen.

Renz M. & Kurz C. (1984) A colorimetric method for DNA hybridization. *Nucleic Acids Research* **12**, 3435–44.

Rigby P.W.J., Dieckmann M., Rhodes C. & Berg P. (1977) Labelling deoxyribonucleic acid to high specific activity *in vitro* by nick translation with DNA polymerase I. *Journal of Molecular Biology* **113**, 237–51.

Saiki R.K., Gelfand D.H., Stoffel S. *et al.* (1988) Primer-directed enzymatic amplification of DNA with a thermostable DNA polymerase. *Science* **239**, 487–91.

Sambrook J., Fritsch E.F. & Maniatis T. (1989) *Molecular Cloning: A Laboratory Manual*, 2nd ed. Cold Spring Harbor Laboratory, Cold Spring Harbor, New York.

Schaad N.W., Azad H., Peet R.C. & Panopoulos N.J. (1989) Identification of *Pseudomonas syringae* pv. *phaseolicola* by a DNA hybridization probe. *Phytopathology* **79**, 903–7.

Schoulties C.L., Civerolo E.L., Miller J.W. *et al.* (1987) Citrus canker in Florida. *Plant Disease* **71**, 388–94.

Stead D.E. & Pemberton A.W. (1987) Recent problems with *Pseudomonas syringae* pv. *pisi* in U.K. *EPPO Bulletin* **17**, 291–4.

Szabo L.J. & Mills D. (1984) Integration and excision of pMC7105 in *Pseudomonas syringae* pv. *phaseolicola*: involvement of repetitive sequences. *Journal of Bacteriology* **157**, 821–7.

van Brunt J. & Klausner A. (1987) Pushing probes to market. *BioTechnology* **5**, 211–21.

Walker J. & Dougan G. (1989) DNA probes: a new role in diagnostic microbiology. *Journal of Applied Bacteriology* **67**, 229–38.

Ward L.J. & De Boer S.H. (1990) A DNA probe specific for serologically diverse strains of *Erwinia carotovora*. *Phytopathology* **80**, 665–9.

Appendix 1: Genomic fingerprinting

The method is based on the procedures of Bradbury *et al.* (1984), Hartung & Civerolo (1987) and Cooksey & Graham (1989).

Materials
- PBS: 20 mM potassium phosphate buffer, pH 6.9, containing 150 mM NaCl.
- 50TE: 50 mM Tris, pH 8.0, containing 50 mM EDTA.
- Egg white lysozyme.
- Lysing solution
  0.5% SDS
  50 mM Tris-HCl, pH 7.5
  400 mM EDTA
  1 mg ml^{-1} pronase.
- Tris buffer-saturated phenol, pH 7.8.
- 10TE: 10 mM Tris-HCl, pH 8.0, containing 1 mM EDTA.
- Restriction endonucleases.
- 3.0 M sodium acetate.
- RNase A: dissolve 10 mg ml^{-1} in 10TE, heat to 100°C for 15 min and allow to cool slowly to room temperature. Store at −20°C in 15 μl aliquots, and dilute 15 μl to 3 ml before use.
- Chloroform.
- Ethanol.

Methods
- Inoculate two 10 ml Luria broth cultures with a single colony per strain. Incubate for 18 h with shaking at 27°C.
- Pool the cultures and centrifuge at 10 000 **g** for 10 min.
- Resuspend the pellet in 10 ml PBS.

Repeat centrifugation. Resuspend the pellet in 5 ml 50TE buffer.

- Add egg white lysozyme to a final concentration of $1 \, mg \, ml^{-1}$. Incubate at 0°C for 30 min.
- Add 1 ml freshly prepared lysing solution.
- Incubate each tube at 50°C until the suspension has cleared.
- Extract the lysate with an equal volume of Tris buffer-saturated phenol.
- Centrifuge at 9000 $\boldsymbol{g}$ for 10 min and transfer the aqueous supernatant to a clean tube.
- Add sodium acetate to a final concentration of 0.3 M.
- Add 2 volumes of ethanol and mix by inversion.
- Use a glass pipette (or rod) to spool the nucleic acids and then dissolve them in 3 ml 10TE buffer containing RNase A ($50 \, \mu g \, ml^{-1}$). Incubate for 30 min at 37°C.
- Extract the resulting solution with an equal volume of chloroform.
- Repeat ethanol precipitation and spool the nucleic acids. Dissolve the resulting DNA in 1 ml 10TE buffer.
- Determine the concentration of DNA ($\mu g \, ml^{-1}$) spectrophotometrically: $OD_{260} \times 50 \times$ dilution.
- Each restriction endonuclease digestion should contain $3-5 \, \mu g$ DNA and enzymes should be used as instructed by the manufacturers. Typical enzyme concentrations would be 20 units in a reaction volume of $35-55 \, \mu l$.
- Samples may be loaded on an agarose gel (for separation of fragments typically $2-20 \, kb$) or a polyacrylamide gel (for separation of fragments $<2 \, kb$ in size). Large ($>20 \, kb$) fragments are best separated by field inversion gel electrophoresis.

Vertical 5%
polyacrylamide gels
- Gel dimensions: 1.5 mm thick, 14 cm long.
- Electrophorese at 14 mA constant current for 14 h.
- Tank buffer: TBE (89 mM Tris HCl; 89 mM boric acid; 2 mM EDTA.).

Horizontal 1%
agarose gels
- Slab gel: $0.5 \times 15 \times 20 \, cm$.
- Electrophorese at 30 V for 16 h.
- Tank buffer: TAE (40 mM Tris HCl; 20 mM sodium acetate; 2 mM EDTA).

For both types of gel, stain and then photograph on a transilluminator. Ethidium bromide can be used at $2 \, \mu g \, ml^{-1}$ with orange and yellow filters with Polaroid type-55 high contrast film, or at $1 \, \mu g \, ml^{-1}$ with red and yellow filters with Polaroid P/N 665 film.

Appendix 2: Colony blotting and hybridization
The method is based on the procedure of Grunstein & Hogness (1975). We use this procedure for many plant pathogenic bacteria.

Materials
- NaOH/NaCl: 0.5 M/1.5 M.
- Tris buffer: 1 M Tris HCl, pH 7.4.
- TBS
    0.5 M Tris HCl, pH 7.4
    1.5 M NaCl.
- $100\times$ conc. Denhardt's solution

> 2% Ficoll (mol. wt 400 000)
> 2% PVP (mol. wt 360 000)
> 2% BSA.

- 10% SDS.
- 20× conc. SSC.

> NaCl   175.3 g
> Sodium citrate   88.2 g

Adjust pH to 7.0 with 10 M NaOH.

- Denatured salmon sperm DNA (Sigma, Type III sodium salt).

Method
- Replica-plate colonies directly from master plates or dilution plates to Hybond-N nylon membrane (Amersham)* by placing the membrane on the surface of an agar plate. Ensure that the membrane is adequately labelled both for identification and orientation (a syringe needle dipped in ink and stabbed through the membrane edge in 3 asymmetric positions will suffice).
- Grow colonies for 36–48 h at 25°C.
- Remove the membrane and place (colony-side up) on a pad of filter paper (previously soaked well in NaOH/NaCl solution). Ensure that no air bubbles are trapped beneath the membrane. This should lyse the colonies and denature the DNA. Do not allow any liquid to flow onto the upper surface of the membrane.
- After about 10 min transfer the membrane to a pad of filter paper that has been well soaked in Tris buffer. Remove the membrane after 2 min and transfer to a fresh pad soaked in Tris buffer for a further 2 min.
- Remove the membrane and place on a pad of filter paper that has been soaked in TBS.
- Remove the membrane after approx. 15 min and air dry until colonies are dry.
- Wrap in Saranwrap (Dow Chemical Co.) and irradiate on an ultraviolet transilluminator (for 1 min or as determined – see manufacturer's instructions for use of nylon membranes).
- Prehybridization: two colony blot filters are placed in a sandwich box containing the following.

> 20× SSC   15 ml
> 100× Denhardt's solution   2.5 ml
> 10% SDS   2.5 ml
> Sterile distilled water   30 ml
> Denatured salmon sperm DNA   100–200 µl

Leave in a shaking water bath at 65°C for 1 h before adding probe.
- Preparation of denatured salmon sperm DNA: dissolve the DNA in water at a concentration of 10 ng ml^{-1}. This may take a little while. Shear the DNA by passing it several times through an 18-gauge hypodermic needle. Boil it for 10 min and store at −20°C in small aliquots (100–200 µl).
- Hybridize the filter with ^{32}P-labelled probe DNA (see Appendix 3).

---

* (See Amersham booklet entitled *Blotting and Hybridization Protocols for Hybond Membranes*).

The probe may consist of a specific cloned fragment, a PCR-amplified sequence or total plasmid DNA from a particular bacterium.

- Washing blots after hybridization:

  Dispose of the contents of the sandwich box in an approved place. Add enough 2× conc. SSC to cover the blot and incubate in a shaking water bath at 65°C for 15 min.

  Pour out the contents and repeat the previous step.

  Pour out the contents. Incubate the filters with 2× conc. SSC containing 0.1% SDS at 65°C for 30 min.

  Air dry filters just enough to get rid of the surface moisture.

  Wrap in Saranwrap.

- Developing:

  In a dark room, lit only by red light, put the colony blots (still wrapped) into a cassette next to X-ray film which has two intensifying screens (Ilford), one on either side, and wrap the whole cassette in a black polythene bag.

  Leave at −70°C for two nights.

  Remove from the freezer and leave at room temperature for 30 min.

- Develop in the dark room as follows. Make up the developer and fixer as outlined by the manufacturer. The developer should be discarded after use but the fixer can be re-used. Take care to handle the film only by the edges. Immerse in developer for 6 min. Wash in tap water for 5−6 min. Immerse in fixer for 6 min. (The light may be switched on after only 2 min.) The plate is then washed under running water for 30 min and dried in a drying cabinet.

Appendix 3: Radio-labelling of probe DNA (nick-translation) and hybridization

The radio-labelling procedure is based on the instructions supplied with a kit available from Gibco-BRL (catalogue no. 81605B).

Materials   *Solution A*: 0.2 mM of each of the deoxynucleoside triphosphates (dNTPs) listed below in 500 mM Tris-HCl, pH 7.8, containing 50 mM MgCl$_2$ and 100 mM 2-mercaptoethanol.

A1: (no dATP) containing dCTP, dGTP, dTTP.
A2: (no dCTP) containing dATP, dGTP, dTTP.
A3: (no dGTP) containing dATP, dCTP, dTTP.
A4: (no dTTP) containing dATP, dCTP, dGTP.
A5: (no dCTP, no dGTP) containing dATP, dTTP.

Solution B: 5 μg phage λ DNA, 0.1 mM EDTA and 120 mM NaCl in 10 mM Tris-HCl, pH 7.5.

Solution C: DNA polymerase I/DNase I (100 units).
0.4 units μl^{-1} DNA polymerase I/40 pg μl^{-1} DNase I in 50 mM Tris-HCl, pH 7.5 containing:

5 mM magnesium acetate
1 mM 2-mercaptoethanol
0.1 mM phenylmethylsulphonyl fluoride (PMSF)

50% (v/v) glycerol

100 µg ml^{-1} nuclease-free BSA.

Solution D: 300 mM EDTA, pH 8.0 (stop buffer).

Solution E: distilled water.

Solution F: 0.1 mM EDTA in 10 mM Tris-HCl, pH 7.5.

Method   To a screw-cap Eppendorf tube on ice add the following.
* 1 µg DNA sample in 20 µl (amount can vary depending upon concentration) solution F or 1 µg λ DNA in solution B.
* 5 µl of solution A2.
* 1 µl of ^{32}P-labelled dCTP (370 KBq µl^{-1}) at activity date, half-life 14 days; double the volume after each half-life).
* Solution E to 45 µl.
* Add 5 µl solution C, mix gently but thoroughly, then spin in a microfuge for 5 s and incubate for 1 h at 15°C.
* Add 5 µl solution D.
* The reaction mixture is passed through a Sephadex G-50 column (Pharmacia) according to the manufacturer's instructions, to separate the probe from unincorporated dNTPs.
* Boil the labelled DNA probe in solution F (to separate the strands of DNA) for 10 min in a dry block at 100°C. Heating does not need to be vigorous.
* Add to the contents of the sandwich box (see Appendix 2: filters which have been prehybridized for 1 h at 65°C) and hybridize overnight at 65°C in a shaking water bath. (Remember to check the water level before leaving overnight.)

Note that the procedure described above assumes use of radioactively labelled dCTP but, of course, any of the other dNTPs could be used as label with appropriate adjustment of solution A.

Appendix 4: Preparation of samples for dot- and squash-blots (Gilbertson *et al.*, 1989)

Dot-blots   Three leaf discs excised from the margins of lesions are crushed in 0.5 ml sterile distilled water in a 1.5 ml Eppendorf tube with a sterile plastic pestle (Kimble–Kontes). Aliquots (10 µl) of the resultant suspension are spotted onto membrane filters for lysis etc., as described in Appendix 2. If desired, diluted suspensions can be plated onto appropriate culture media to determine the viable count of organisms present.

Squash-blots   Leaf discs are excised from the margins of lesions with a number 4 cork borer. The discs are placed on a nylon membrane filter and squashed with a sterile, round-bottomed glass rod. For lysis of bacterial cells etc. see Appendix 2.

9 Use of RFLPs to identify races of fungal pathogens

A. CODDINGTON & D.S. GOULD

Introduction

The identification of fungal pathogens is traditionally based on morphology but the techniques of molecular biology are being increasingly applied to difficult problems of identification, sometimes at the species level but more often at the level of subspecies, varieties, formae speciales and races. At each of the levels mentioned there are fewer and fewer morphological characteristics which can be used to distinguish between isolates of a pathogen. Subspecies and varieties differ in some aspects of morphology but formae speciales and races normally differ only in host range; formae speciales attack different host species and races attack cultivars of the same host species.

It is generally accepted that new races arise in the field as a means of overcoming resistance genes in the host plant. Once resistance has been overcome in this way the plant breeder has to introduce new resistance genes in the host to prevent infection by the new race of the pathogen. For new isolates of fungal pathogens which show such cultivar specificity it is of practical importance to establish the race to which it belongs. At present, races can only be defined with respect to known resistance genes in the host. In principle a series of host differentials, homozygous for different combinations of resistance genes and preferably in an isogenic background, is infected with the new race isolate and the host plants are scored for resistance or susceptibility. From the resulting pattern obtained the race can be defined. An example involving the four races of *Fusarium oxysporum* f.sp. *pisi* is shown in Table 9.1.

In practical terms it is a time-consuming business to carry out such a race identification on a new isolate. Scoring of symptoms is often subjective and the infection process, which depends on temperature, light intensity, humidity, etc., can be quite variable even with the same combination of race and cultivar (see Coddington et al. (1987) for a fuller discussion of the problem).

To overcome the problems associated with the standard method for race classification we have attempted to make use

Table 9.1 Classification of race isolates of *Fusarium oxysporum* f.sp. *pisi* by the resistance reactions produced on the six host differentials of Haglund and Kraft (1979).

Fungal race	Host differential					
	New Season	New Era	Dark Skin Perfection	WSU28	WSU23	Little Marvel
1	R	R	R	R	R	S
2	R	R	S	S	R	S
5	S	S	S	R	R	S
6	R	S	S	R	S	S

R = resistant; S = susceptible reactions.

of the natural variation present in the DNA of all species as a means for classifying the races of a fungal pathogen. The way we do this is to look for restriction fragment length polymorphisms (RFLPs). Restriction endonucleases are enzymes which cut DNA at a particular sequence of four to six bases. An example of a restriction enzyme is *Eco*R I which cuts DNA at the sequence GAATTC whenever it occurs. The fragmented DNA is separated by agarose gel electrophoresis and the fragments are made visible by staining with a dye (ethidium bromide) or by hybridizing with a radio-labelled probe and autoradiography. Agarose gel electrophoresis separates DNA molecules according to size: the smaller the fragment the further the distance it moves in a given time. The genome of a fungal pathogen such as *Fusarium* contains about 3.0×10^7 base pairs (bp) or 30 000 kilobase pairs (kb) of DNA. Statistically, a sequence such as GAATTC occurs once every 4000 bp so, if the DNA of the whole genome was cut with *Eco*R I, one would expect to get about 7500 fragments, all of different size since the position of the sequence along the DNA molecule is also random. This would mean that no individual DNA fragment would be present in sufficient quantity to be visible by the above staining technique against a background of all the other fragments. However, about 20% of the *Fusarium* genome consists of relatively short sequences of DNA repeated many times. This DNA is mainly rDNA (the DNA coding for ribosomal RNA) and mtDNA (the DNA from mitochondria), of which there are many copies per cell. When this repetitive DNA is cut with *Eco*R I some fragments will be present many hundreds of times and will be visible after electrophoresis and staining. This is shown clearly in Fig. 9.1 where the repetitive DNA is seen as white bands on a grey background of single-copy DNA.

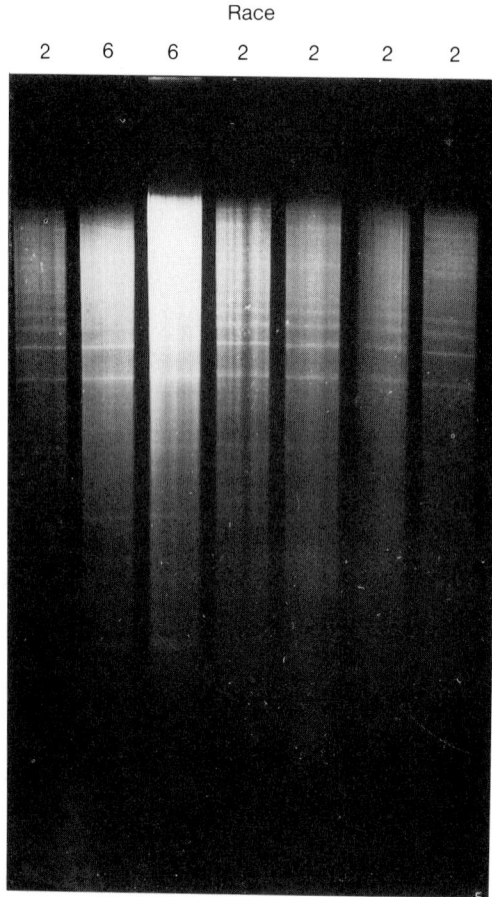

Fig. 9.1 Restriction enzyme digestion patterns of total DNA from seven race isolates of *F. oxysporum* f.sp. *pisi*. Total DNA (1 μg) from five isolates of race 2 and two isolates of race 6 was digested with *Eco*R I. The resulting fragments were separated on an agarose gel and stained with ethidium bromide.

The repetitive DNA banding pattern so obtained is clearly a characteristic of the race.

Natural variation in a DNA molecule most commonly involves either the change of one base pair for another or the insertion or deletion of a single base pair or longer piece of DNA. Such changes can lead to the removal or addition of a restriction site and also affect the relative spacing between two restriction sites. In both cases differences in the length of a restriction fragment would be generated; these show up as an alteration in its position after electrophoresis on an agarose gel. If variation occurred in each unit of a repeated segment of DNA, by the mechanism of gene conversion for example, then this would readily be seen. Hence, if two races differ in the sequences of the repetitive DNA one could detect this rather easily and this is the case for the races of *F. oxysporum* f.sp. *pisi* (see Figs 9.2 & 9.3). Figure 9.3 shows restriction enzyme digests of mtDNA obtained with two different enzymes; for each enzyme the banding pattern is clearly different between races. A comparison betwen Figs 9.1 and 9.3 (*Eco*R I tracks) allows the bands in Fig. 9.1, originating from mtDNA in race 2, to be located.

However, what if the variation does not occur in the repetitive DNA but in the other 80% of the DNA which has so-called unique sequences? This DNA represents the coding capacity of the genome for proteins. In this case differences in fragment lengths must be detected by Southern blotting and hybridization to a radioactively labelled piece of DNA used as a probe.

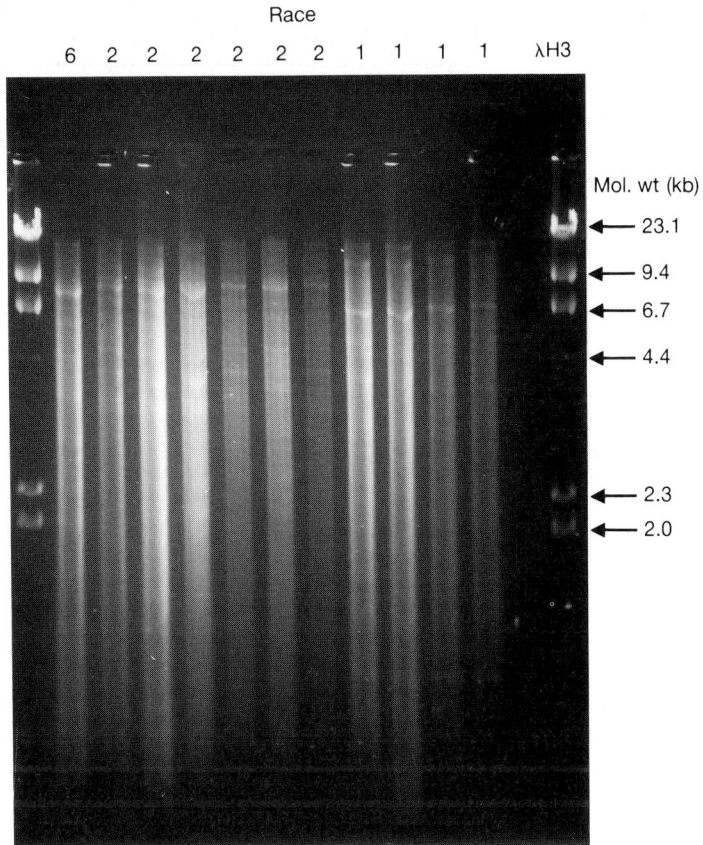

Fig. 9.2 Restriction enzyme digestion patterns of total DNA from three different races of *F. oxysporum* f.sp. *pisi* using two enzymes. Total DNA (1 µg each) from four isolates of race 1, six isolates of race 2 and one isolate of race 6 was digested with a mixture of *Xba* I and *Xho* I. The resulting fragments were separated on an agarose gel and stained with ethidium bromide. Molecular weight markers are *Hin*d III digests of phage lamda (λH3).

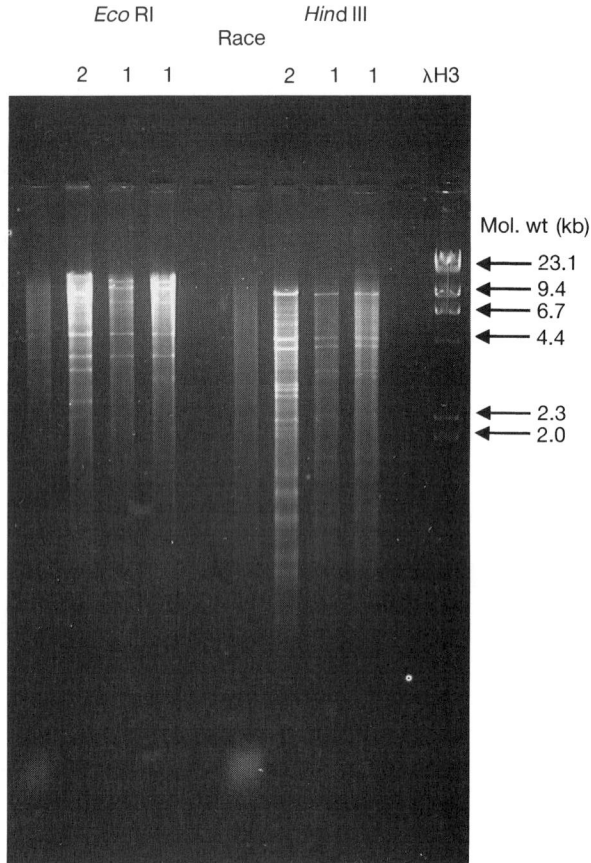

Fig. 9.3 Restriction enzyme digestion patterns of mitochondrial DNA from two different races of *F. oxysporum* f.sp. *pisi*. Total DNA (1 μg each) from two isolates of race 1 and one isolate of race 2 was digested separately with *Eco*R I and *Hin*d III. The resulting fragments were separated on an agarose gel and stained with ethidium bromide. (λH3 markers are the same as in Fig. 9.2).

Again, a change in the restriction site pattern is detected by an alteration in the length of a fragment which, in turn, alters its position after agarose gel electrophoresis. However, in this case only a small region of DNA is highlighted on the gel. An example of the technique is given in Fig. 9.4 in which total DNA from different races of *F. oxysporum* f. sp. *pisi* were digested with two restriction enzymes and probed with ^{32}P-labelled pRE1, which contains a flax rDNA repeat (Goldsborough & Cullis, 1981). The DNA fragments shown in Fig. 9.2 were first transferred to a nylon membrane by Southern blotting before being probed. A simple banding pattern was obtained since only the DNA fragments homologous to the radioactive probe show up by this method. Again, it can be seen that the banding pattern is a characteristic of the race.

Race

6   2   2   2   2   2   2   1   1   1   1

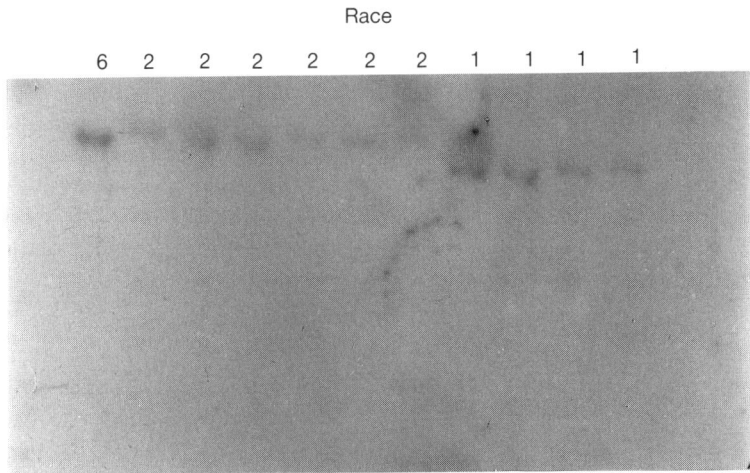

Fig. 9.4 Total DNA from different races of *F. oxysporum* f.sp. *pisi* digested with a mixture of *Xba* I and *Xho* I and probed with ^{32}P-labelled pREI.

Strategy for obtaining a race-specific RFLP

The strategy given below is based upon the approach we have taken in our work on *F. oxysporum* f. sp. *pisi*. This pea root pathogen grows readily in a defined medium and there are no problems in obtaining enough starting material for DNA isolation. The same strategy could be used for any fungal pathogen which will grow in axenic culture and has been used in our department for *Fulvia fulva* (syn. *Cladosporium fulvum*), *Aschochyta pisi*, *Mycosphaerella pinodes* and *Leptosphaeria maculans*. If the fungal pathogen cannot be grown in axenic culture it is possible to start with a spore or conidial suspension. DNA has even been extracted from infected leaves or from spores harvested from leaves infected with the obligate pathogens *Erysiphe graminis* (O'Dell *et al.*, 1989; Brown *et al.*, 1990) and *Bremia lactucae* (Hulbert & Michelmore, 1988) and used for RFLP studies.

The strategy, applied here to identify races, can also be used at other taxonomic levels to separate species, subspecies, varieties and formae speciales. The number of RFLPs which could prove useful for the purpose of identification at these other levels may be greater than at the level of races. As the techniques are very similar at whatever level is being investigated we will only refer to them as applied to the races of *F. oxysporum* f. sp. *pisi*.

Outline of strategy   1 Production of fungal mycelium:
(a) isolate from stem section of an infected plant;

(b) grow mycelium in liquid culture;

(c) Harvest and freeze dry.

2 DNA extraction.

3 Restriction enzyme digestion.

4 Agarose gel electrophoresis, staining and visualization of repetitive DNA bands.

5 Fractionation of total DNA into rDNA, mtDNA and non-repetitive nuclear DNA on a caesium chloride (CsCl)−bisbenzimide gradient; then repeat steps 3 and 4 on the rDNA and mtDNA fractions.

6 Probing the total DNA with a radioactive probe.

(a) After step 4 transfer the fragments to a nylon or nitro-cellulose filter (Southern blotting).

(b) Prepare the radioactive probe.

(c) Hybridize to the filter, wash and dry.

(d) Visualize radioactive bands by autoradiography.

RFLPs in the repetitive DNA should be seen after step 4 but more conclusive results can often be seen with the purified repetitive DNA after step 5. If the RFLPs are in the non-repetitive nuclear DNA then they are generally much harder to find and one would have to try many different restriction enzymes and probes before being successful. The simplest probes to try are usually those of cloned DNA from other species which have already shown variation in that species, e.g. the flax rDNA repeat pRE1 (Goldsborough & Cullis, 1981).

Methods

General   The methods described below are routinely used in our labor-atory. For those new to molecular biology the possession of a practical manual such as *Molecular Cloning, A Laboratory Manual* (Sambrook *et al.*, 1989) is essential, and all the basic techniques, equipment and procedures are described therein. A good theoretical introduction is *Gene Cloning, An Introduction* (Brown, 1990). Materials, equipment and media can be obtained from Oxoid or Difco, restriction enzymes from Gibco-BRL or Pharmacia and biochemicals from Sigma or BCL. Other chemi-cals should be as pure as possible. A high-speed bench centrifuge (Eppendorf or MSE microfuge, supplied by BDH and Fisons respectively), a general-purpose refrigerated centrifuge (DuPont-Sorvall) and a high-speed preparative centrifuge (Beckman) are also required.

Production of fungal mycelium   Any defined liquid medium which supports good growth of all isolates of the pathogen under investigation could be used. For *F. oxysporum* f. sp. *pisi*, the fungus is first isolated from infected pea stems by plating out surface-sterilized stems on to potato dextrose agar (Appendix 1) and then incubating the plates at 25°C for about 2 days. Liquid medium CDAZ (Appendix 1) (100 ml) is then inoculated with a small plug of agar taken from the edge of the colony and incubated at 25°C in the dark for 5−7 days.

Mycelia are harvested by filtration on to Whatman 3 mm filter paper and then freeze-dried. Freeze-drying is not essential but it makes subsequent grinding easier.

Extraction of fungal DNA Fungal DNA is extracted by a method adapted from Raeder & Broda (1985). A protocol for the extraction of fungal DNA from freeze-dried mycelium is given in Appendix 2. As efficient extraction depends on the fineness of the ground particles, it is essential that grinding is both thorough and complete.

After extraction, a small amount of the DNA solution (2−5 µl) is run on a gel with standard amounts of fragment size markers (λ *Hin*d III). The concentration can be estimated after visualization under ultraviolet light.

DNA restriction enzyme digestion Approximately 2 µg of the DNA is placed in an Eppendorf centrifuge tube with 2 µl of the appropriate digestion buffer (see manufacturer's instructions) and 10 units of enzyme. The volume is made up to 20 µl with sterile distilled water and the sample centrifuged in a microcentrifuge for 1 s to ensure that the contents are well mixed; it is then incubated at 37°C for 6−7 h. The amount of restriction enzyme is slightly in excess to that given by the manufacturers (1 unit µg^{-1} DNA). This is to compensate for the inhibitory effect of any contaminants in the DNA. After treatment the mixture may be stored at −20°C.

Gel electrophoresis In our laboratory agarose gel electrophoresis is normally run overnight at 20 v (constant voltage) or until the dye incorporated with the digest reaches the end of the gel. Ethidium bromide is incorporated into the gels so that the bands of DNA can be seen under ultraviolet light. Full details of the reagents and the techniques are given in Appendix 3.

CsCl− bisbenzimide gradients for DNA fractionation This protocol is taken from Garber & Yoder (1983) and adapted for 16 × 76 mm tubes and a Beckman type-65 rotor. CsCl (1.15 g ml^{-1}) is dissolved in a DNA solution (a maximum of 1 mg of DNA), bisbenzimide is added at 120 µg ml^{-1} and the tube is filled to just below the level of the cap with 1.15 g ml^{-1} CsCl solution. The tube is filled completely by the addition of liquid paraffin and is then sealed, ensuring that no air bubbles are trapped. It is centrifuged at 190 000 × g for 24 h. When viewed under ultraviolet light the DNA appears as three discrete bands. The uppermost band is mtDNA, the middle is nuclear DNA and the lowest is rDNA. The bands can be removed separately using a syringe inserted through the top of the tube. It may be necessary to repeat the process to further purify the sample. Bisbenzimide is removed by several extractions with isopropanol (removing and discarding the upper layer). The DNA can then be precipitated and digested with enzymes as described above.

Probing the total DNA with a radioactive probe   Once the DNA fragments have been separated in a gel they can be transferred by Southern blotting to a nylon or nitro-cellulose membrane and then fixed (Appendix 4). The filters can be stored between Whatman 3MM paper at room tempera-ture until required. Hybridization and autoradiography are described in detail in Appendix 5.

Conclusions and future prospects

RFLPs have been used in recent years to investigate phylo-genetic relationships of plant pathogenic fungi at various taxonomic levels in a range of phlya including Oomycetes, Ascomycetes and Basidiomycetes.

There have been several studies in which RFLPs have been used to distinguish closely related fungal species from one another. Kohn *et al.* (1988), by probing DNA enzyme digests of seven *Sclerotinia* spp. with clones of *Neurospora crassa* DNA, found a number of RFLPs in rDNA and mtDNA which were useful in separating species. Each species had one to four frag-ments in its mtDNA which were unique and constant within that species. They confirmed previous suspicions that *S. ficariae* was synonymous with *S. sclerotinia*. Similar investigations into taxonomically difficult genera such as *Armillaria* and *Rhizo-ctonia* have yielded new insights into interspecific relationships in these genera (Anderson *et al.*, 1989; Vilgalys & Gonzalez, 1990).

In *Phytophthora*, where there is a dearth of good morphological characteristics, Forster *et al.* (1989) used RFLPs of mtDNA to show that *P. megasperma* f.sp. *glycinea* and *P. megasperma* f.sp. *medicaginis* are genetically distinct and represent different biological species within *P. megasperma*. Differences have also been found between the mtDNAs of formae speciales of *F. oxysporum* (Kistler & Benny, 1989).

Although one might expect fewer differences between the various races of a single species or forma specialis than between species, it is clear from the results reported in Figs 9.1 and 9.2 that RFLPs in the repetitive DNA of *F. oxysporum* f.sp. *pisi* can allow us to distinguish easily between races 1, 2 and 6. Digestion of isolated mtDNA (Fig. 9.3) again shows race specific differences and allows us to define the repetitive DNA bands in the total DNA enzyme digests which are of mitochondrial origin. Probing with a heterologous flax rDNA probe also distinguishes between the three races 1, 2 and 6 (Fig. 9.4).

How universally applicable are these methods to races of other fungal pathogens? RFLPs in repetitive DNA generated by the restriction enzyme *Bam*H I can distinguish one of the five pathotypes of *Ascochyta pisi* (BP1) from the rest, and probing the same digest with pRE1 distinguishes a different pathotype (BP4) from the others. However, when the DNA from the five races of *Mycosphaerella pinodes* was investigated by the same methods no clear cut differences were seen — there was as much variation within race isolates as between them (S. Clulow, pers. comm.). The tomato leaf mould pathogen *Fulvia fulva* has seven races, but no variation is observed when one looks at total DNA digests or on probing with the flax DNA probe pRE1. However, some variation can be seen in the mtDNA and on using a homologous DNA probe from a genomic DNA library of *Fulvia fulva* (N.J. Talbot, pers. comm.). If one follows the strategy outlined at the beginning of the chapter and tries as many restriction enzymes and probes as possible, then eventually variation should be found. However, only variation at the repetitive DNA level which can be detected by restriction digestion of total DNA is likely to be of use in routine screening programmes.

Probing techniques would probably be too expensive and complicated for widespread use and are therefore only briefly outlined here. An interesting variant is the use of human hypervariable minisatellite probes. Using this technique Braithwaite and Manners (1989) detected RFLPs in different subgroups of *Colletotrichum gloeosporioides* and were able to distinguish the two forms of this fungus which cause anthracnose of *Stylosanthes* spp. in Australia.

While the RFLP techniques described here will undoubtedly retain an important place in studies of races and other taxa of plant pathogenic fungi, it is likely that future studies will make increasing use of amplification techniques such as the polymerase chain reaction (PCR) (Vivian, this volume). Use of random primers with PCR might reveal differences between races at the nucleic acid level. If polymorphisms could be detected by this method then the testing of isolates for epidemiological and population genetical studies could be done much more rapidly and without need for radio-labelled probes.

Acknowledgements

We would like to thank S. Clulow and N. Talbot for many stimulating discussions and for allowing us to quote their work on other fungal pathogens prior to publication. D.S. Gould is the holder of a SERC studentship.

References

Anderson J.B., Bailey S.S. & Pukkila P.J. (1989) Variation in ribosomal DNA among biological species of *Armillaria*, a genus of root-infecting fungi. *Evolution* **43**, 1652–62.

Braithwaite K.S. & Manners J.M. (1989) Human hypervariable minisatellite probes detect DNA polymorphisms in the fungus *Colletotrichum gloeosporioides*. *Current Genetics* **16**, 473–5.

Brown J.K.M., O'Dell M., Simpson C.G. & Wolfe M.S. (1990) The use of DNA polymorphisms to test hypotheses about a population of *Erysiphe graminis* f.sp. *hordei*. *Plant Pathology* **39**, 376–90.

Brown T.A. (1990) *Gene Cloning, An Introduction*, 2nd edn. Van Nostrand, London.

Coddington A., Matthews P.M., Cullis C. & Smith K.H. (1987) Restriction digest patterns of total DNA from different races of *Fusarium oxysporum* f.sp. *pisi* — an improved method for race classification. *Journal of Phytopathology* **118**, 9–20.

Feinberg A.P. & Vogelstein B. (1983) A technique for radiolabelling DNA restriction fragments to a high specific activity. *Analytical Biochemistry* **132**, 6–13.

Forster H., Kinscherf T.G., Leong S.A. & Maxwell D.P. (1989) Restriction fragment length polymorphisms of the mitochondrial DNA of *Phytophthora megasperma* isolated from soybean, alfalfa and fruit trees. *Canadian Journal of Botany* **67**, 529–37.

Garber R.C. & Yoder O.C. (1983) Isolation of DNA from filamentous fungi and separation into nuclear, mitochondrial , ribosomal and plasmid components. *Analytical Biochemistry* **135**, 416–22.

Goldsborough P.B. & Cullis C.A. (1981) Characterization of the genes for ribosomal RNA in flax. *Nucleic Acids Research* **9**, 1301–9.

Haglund W.A. & Kraft J.M. (1979) *Fusarium oxysporum* f.sp. *pisi* race 6: occurrence and distribution. *Phytopathology* **69**, 818–20.

Hulbert S.H. & Michelmore R.W. (1988) Restriction fragment length polymorphism and somatic variation in the lettuce downy mildew fungus, *Bremia lactucae. Molecular Plant–Microbe Interactions* **1**, 17–24.

Kistler H.C. & Benny U. (1989) The mitochondrial genome of *Fusarium oxysporum. Plasmid* **22**, 86–9.

Kohn L.M., Petsche D.M., Bailey S.R., Novak L.A. & Anderson J.B. (1988) Restriction fragment length polymorphisms in the nuclear and mitochondrial DNA of *Sclerotinia* species. *Phytopathology* **78**, 1047–51.

O'Dell M., Wolfe M.S., Flavell R.B., Simpson C.G. & Summers R.W. (1989) Molecular variation in populations of *Erysiphe graminis* on barley, oats and rye. *Plant Pathology* **38**, 340–51.

Raeder U. & Broda P. (1985) Rapid preparation of DNA from filamentous fungi. *Letters in Applied Microbiology* **1**, 17–20.

Rigby P., Dieckman M., Rhoades C. & Berg P. (1977) Labelling deoxyribonucleic acid to a high specific activity *in vitro* by nick translation with DNA polymerase I. *Nucleic Acids Research* **1**, 1263–81.

Sambrook J., Fritsch E.F. & Maniatis T. (1989) *Molecular Cloning: A Laboratory Manual*, 2nd edn. Cold Spring Harbor Laboratory, Cold Spring Harbor, New York.

Vilgalys R. & Gonzalez D. (1990) Ribosomal DNA restriction length poly-morphisms in *Rhizoctonia solani*. *Phytopathology* **80**, 151–8.

Appendix 1: Media for the production of mycelium of *Fusarium oxsporum* f.s.p. *pisi*

- Potato dextrose agar (per litre)

Potato dextrose broth (Difco)	24 g
Agar	15 g.

- AZ mineral salts solution (per litre)

$CuSO_4.5H_2O$	22 mg
$MnCl_2.4H_2O$	0.1 g
$ZnCl_2$	0.1 g
$Ca(NO_3)_2.6H_2O$	0.1 g
$BaCl_2.2H_2O$	20 mg
$(NH_4)6Mo_7O_{24}.4H_2O$	20 mg.

- Czapek Dox plus AZ liquid medium (CDAZ) (per litre)

Czapek Dox salts (Oxoid)	33.4 g
AZ solution	10 ml.

Appendix 2: Extraction of fungal DNA

Materials
- Extraction buffer

  0.5 M NaCl

  10 mM Tris-HCl, pH 7.5

  10 mM EDTA

  1% SDS.

- Phenol. This can be purchased ready for use or prepared in the laboratory from crystalline phenol.

  Crystalline phenol is redistilled at 160°C to remove contaminants. Observe all safety precautions when distilling phenol).

  8-hydroxyquinoline is added to a final concentration of 0.1% after distillation to prevent oxidation.

  The liquid phenol is then extracted several times with an equal volume of 1M Tris, pH 8.0, until the pH of the aqueous phase is >7.6. Store under 0.1M Tris, pH 8.0, at 4°C.

- Chloroform: isoamyl alcohol (24:, v/v).
- Sodium acetate solution (3M adjusted to pH 4.8 with glacial acetic acid).
- RNase.
- TE: 10 mM Tris-HCl, pH 7.5; 1mM EDTA, pH 8.
- Ethanol.

Method
- Grind freeze-dried mycelium (0.5 g) in a pestle and mortar with liquid nitrogen. The efficiency of DNA extraction depends greatly on the fineness to which the particles are ground.
- Transfer the powder to a mortar at room temperature, add 2 ml extraction buffer and mix gently with the pestle.
- Add 1.5 ml phenol and mix gently.
- Add 1.5 ml chloroform:isoamyl alcohol and transfer the mixture to a 15 ml Corex tube.

- Spin for 1 h at 12 000 × *g* in an SS-34 rotor in a Sorvall centrifuge. This long spin will sediment high molecular weight polysaccharides.
- Remove the upper, aqueous layer using a wide-bore pipette to avoid shearing the DNA (a disposable Pasteur pipette with the fine tip snapped off is convenient) and transfer to a fresh Corex tube, being careful to avoid the material at the interface.
- Add 25 µl RNase (10 mg ml^{-1}) and incubate at 37°C for 10 min.
- Add an equal volume of phenol and mix gently to form an emulsion.
- Centrifuge at 10 000 rpm (12 000 × *g*) for 10 min.
- Remove the aqueous layer as before and repeat the process at least twice or until the interface clears considerably.
- Add chloroform:isoamyl alcohol and treat as for phenol extraction.

    Phenol and chloroform treatments denature and precipitate proteins and so eliminate fungal nucleases.
- Precipate the DNA by adding 0.1 volume sodium acetate and 2.5 volumes ice-cold 95% ethanol.
- Incubate at −20°C for at least 2 h and then centrifuge at 10 000 rpm (12 000 × *g*) for 20 min.
- Wash the pellet in 70% ethanol and resuspend in a minimal volume of TE (e.g. 250 µl).

    After extraction a small amount of the DNA solution (2−5 µl) is run on a gel with 1 µg λ *Hin*d III fragment size markers to estimate the concentration after visualization under ultraviolet light (the 23 kb band of λ *Hin*d III will represent 0.5 µg).

Appendix 3: Gel electrophoresis

Materials
- TAE (50× conc.)

Tris	242 g
Glacial acetic acid	57.1 ml
0.5 M EDTA, pH 8.0	100 ml

    Make up to 1 litre.
- TBE (5× conc.)

Tris	54 g
Boric acid	27.5 g
0.5 M EDTA, pH 8.0	20 ml

    Make up to 1 litre.
- Loading dye

    0.25% bromophenol blue

    40% (w/v) sucrose

    10 mM EDTA.
- Agarose.
- Ethidium bromide stock solution: 10 mg ml^{-1}.

    Warning: Ethidium bromide is carcinogenic and latex gloves should be worn when handling this material and solutions containing it.

Method
- We use 14 × 11 cm gels in a Gibco-BRL-H5 gel rig with a 14 well comb having teeth of 2 mm thickness.
- TAE or TBE buffer can be used with similar results.
- Dissolve agarose by heating in either 1× conc. TAE or TBE to give

0.8% gel, 80–100 ml (a 100 ml gel is easier to handle if the gel is to be blotted).

- Add ethidium bromide stock solution to a final concentration of $0.5\,\mu g\,ml^{-1}$ and pour the agarose solution (when cool enough to handle) into the gel mould (whose ends have been sealed with tape) with the comb in place.
- Once set, remove the tape and place the gel and mould in the rig with the gel just submerged in buffer. Remove the comb carefully and wash out the wells with buffer.
- Add loading dye to each sample. This makes the solution visible and more dense than the buffer allowing it to be placed more easily in the wells under the buffer.
- Gels are normally run overnight at 20 V (constant voltage) or until the dye reaches the end of the gel.
- The DNA in the gel is visualized by placing it on an ultraviolet transilluminator. The ethidium bromide intercalates with the DNA which then fluorescence blue–white under ultraviolet light.
- Photograph the gel.

Appendix 4: Southern blotting

Materials
- Denaturing solution

NaOH	100 g
NaCl	438 g

Make up to 5 litres.
- Neutralizing solution

Tris	121 g
NaCl	87.8 g

Make up to 1 litre and adjust to pH 8.0 with HCl (about 80 mls).
- Transfer buffer

CH_3COONH_4	385 g
NaOH	4 g

Make up to 5 litres.

Methods
- Place the gel in a shallow tray.
- Wash for 1 h in denaturing solution and then for 1 h in neutralizing solution. The solutions are changed half way through each wash.
- Wrap a piece of Whatman 3 MM paper around an upturned gel mould to act as a wick and place in a shallow dish.
- Place two pieces of 3 MM paper the same size as the gel on top.
- Wet thoroughly and fill the dish with transfer buffer to half way up the gel mould.
- Rinse the gel in transfer buffer, invert it and place it on the 3 MM paper ensuring that no air bubbles are trapped.
- Cut a piece of nylon or nitrocellulose membrane the same size as the gel (wear gloves when handling), wet it in transfer buffer and then lay it on top of the gel, again ensuring that no air bubbles are trapped.
- On top of this place three gel-sized pieces of 3 MM paper soaked in transfer buffer, three dry pieces of 3 MM paper, a small stack of

tissues (5–6 cm high) and then a small plate of glass. A weight (500 g) is finally placed on top.
- To prevent the transfer solution from 'short-circuiting' the gel, place cling film around the edges of the gel and stretch it to wrap around the dish.
- Transfer of DNA is allowed to proceed overnight.
- Fix the DNA onto nylon membrane by placing the filter on an ultraviolet transilluminator for 5 min, or onto nitrocellulose membrane by baking for 2 h at 80°C under vacuum.
- Store filters between 3 MM paper at room temperature.

Appendix 5: Hybridization of ^{32}P-labelled DNA probes

The probe can be labelled by two methods: nick-translation (Salazar & Querci, Vivian, this volume); or multiprime labelling (Rigby *et al.*, 1977; Feinberg & Vogelstein, 1983; Sambrook *et al.*, 1989).

Materials
- SSC (20 × conc.)

NaCl	175.3 g
Sodium citrate	88.2 g

  Make up to 1 litre and adjust to pH 7.0 with 10 M NaOH.
- Denhardt's solution (100× conc.)

BSA	2% (w/v)
Ficoll (mol. wt 400 000)	2% (w/v)
Polyvinylpyrrolidone (mol. wt 360 000)	2% (w/v)

- Prehybridization solution: (6× conc. SSC, 5× conc. Denhardt's solution, 0.5% SDS).

Methods
- Prehybridize the filters overnight as follows.
  Denature sheared herring sperm DNA (0.5 ml of a 10 mg ml^{-1} solution) by heating to 100°C for 5 min and then chilling rapidly on ice.
  Add to 15 ml pre-hydridization solution.
  Pour the solution into a heat-sealable bag containing the filter, seal the bag, expelling all air bubbles, and leave it at 65°C in a shaking water bath overnight.
- Denature the labelled probe if double-stranded by heating at 100°C for 5 min and chilling on ice before adding it to the bag with 0.5 ml herring sperm DNA, denatured as described above.
- Allow it to hybridize to the filter at 65°C overnight.
- High stringency washing is then performed on the filter as follows.
  Incubate for 1 h at 65°C with 2× conc. SSC, 0.1% SDS.
  Incubate for 1 h at 65°C with 0.1× conc. SSC, 0.1% SDS.
We use a high stringency wash when probing with an rDNA repeat that shows a high degree of homology to *Fusarium* DNA. For heterologous probes the temperature is lowered and the concentration of SSC increased.
- Dry the filter (unless it is to be reprobed) and wrap it in cling film or seal it in a bag.
- Autoradiograph.

Section 3
Commercial Aspects and
Practical Applications

10 Screening for pathogens and contaminating micro-organisms in micropropagation

A.C. CASSELLS

Introduction

A perceived advantage of micropropagated plants is their disease-free status; however, this cannot be achieved without rigorous quality control during production. The presence of contaminating micro-organisms in culture can cause direct losses and, as has been shown recently, affect growth rates *in vitro* (Long *et al.*, 1988; Leifert *et al.*, 1989 b). Thus, contamination, in the broad sense, may result in loss of production, reduced productivity and consequential losses if latent pathogenic micro-organisms are expressed in the progeny plants (Hayward, 1974).

In addressing the problem of contamination, it is important to know the source. In micropropagation three primary sources can be distinguished.

1 Endophytic contaminants occurring intracellularly and intercellularly in the explant tissues.

2 Surface contaminants not readily amenable to surface sterilants.

3 Contaminants introduced through faulty techniques. Some of the latter may be associated with the production workers (Leifert *et al.*, 1989 a).

Careful visual examination of cultures will detect contaminants capable of growing on the plant tissue culture media. However, many potentially cultivable organisms may be suppressed on the high-salt, high-sucrose media used particularly in the early stages of tissue culture. These may spread through the tissue in the clonal multiplication phase to be expressed at rooting of microcuttings or in the progeny plants with significant losses.

Arguably more problematic are the subliminal contaminants defined pragmatically as those organisms whose presence cannot be visibly detected but which systemically contaminate the cultures. These include specific pathogens of the crop but also promiscuous endophytes and possibly intransigent surface contaminants; also included are non-cultivable, and intractable, potentially cultivable organisms. The latter may be detected by subculture on to appropriate media, whereas the former can only be detected by specific tests. These organisms may affect productivity or be expressed in progeny, e.g. on transfer to different geographical regions (see Hayward, 1974; Cassells, 1986).

In this contribution, screening and detection of diseases and contaminants in tissue cultures will be discussed in the context of developing practical quality control and assurance standards for the micropropagation industry. A more comprehensive review of quality control in micropropagation is presented elsewhere (Cassells, 1991). Bastiaens (1983) has reviewed the implications for micropropagators of endophytic bacteria in donor plants.

In discussing approaches to the screening of cultures for contaminating micro-organisms, it is useful to refer to the culture stage. While all cultures may be examined at the start of a culture cycle, this is not practical for the multiplication and subsequent phases.

Murashige (1974) divided micropropagation into three stages.
1 Establishment of the tissue cultures.
2 Multiplication phase.
3 Preparation for transfer of tissue culture progeny to the open environment.

The latter stage may involve *in vitro* rooting. All three stages may involve exposure to substantially different media (Debergh & Maene, 1981) which, in turn, may differentially influence the expression of cultivable contaminants. Media effects on non-cultivable contaminants are largely unrecorded.

Visual detection of contaminants

In stage 1 or throughout small-scale production it may be feasible to examine each culture for signs of contamination. Cultivable micro-organisms including bacteria, yeasts and other fungi can be readily observed, the distribution of the growth indicating its possible origin. Close examination of the immediate area of the explant may allow detection of less adapted micro-organisms which are not capable of colonizing the culture medium. These frequently cause a 'halo' to form around the explant.

Numerous authors have advocated the inclusion of bacterial nutrients in the stage 1 culture medium (e.g. Menard *et al.*, 1985) to encourage bacterial expression while the use of clear gels, e.g. 'Gelrite', may facilitate detection.

In some cases cultures in stage 2 or later may show disease symptoms. Debergh and Vanderschaeghe (1988), for example, have observed brown spots on the petioles and leaves *in vitro* of bacterially contaminated gerberas.

In mass micropropagation (stage 2 and later) it is not practical to examine each culture for symptoms or low-level expression of contaminants and so sampling must be employed. Here caution should be exercised as contamination may not be random in distribution.

Culture indexing

A wide range of potentially cultivable micro-organisms has been associated with micropropagated plants (Cassells, 1986; Leifert *et al.*, 1989 a) (Table 10.1). These include plant pathogenic bacteria, bacteria common in soil and on plant surfaces, airborne bacteria and bacteria associated with production workers. The aim of culture indexing is to detect such subliminal contamination by subculture of the organisms concerned on appropriate bacteriological media. With regard to the wide range of bacterial contaminants which have been reported (Table 10.1; see also Campbell, 1985), the choice of media for culture indexing is arbitrarily divided into general or specific (Schaad, 1980; Lelliott & Stead, 1987). Fortuitously, many plant-associated bacteria grow on a variety of relatively simple formulations, as opposed to the less common fastidious intracellular organisms, e.g. mycoplasmas. The latter, like viruses, may be more conveniently detected by special tests (Clark, this volume).

Caution should be exercised in carrying out culture indexing, firstly, in sampling the culture and secondly, in preparing the inoculum. It is implicit that culture indexing is non-destructive of the culture and consequently it must be ensured that the sample taken is not an 'escape'. Dimock (1962) has proposed a model where a middle section is cultured and the upper and lower sections are culture indexed. Regarding the inoculum, care must be taken to ensure that plant factors, e.g. tannins, do not inactivate any bacteria in the tissue extract (Reuther, 1985). In this laboratory we routinely transfer suspect explant tissue intact to bacteriological media for culture indexing. Further, it should be remembered that plant-inhabiting bacteria may have a lower temperature optimum than bacteria from other sources and cultures should be incubated at appropriate temperatures.

Culture indexing may also be employed for the detection of mollicutes (Bove, 1988; Debergh & Vanderschaeghe, 1988) and for other fastidious prokaryotes (Roberts & Ambler, 1988) but, in the case of intracellular micro-organisms *in situ*, staining may be simpler.

Table 10.1 Bacterial contaminants isolated from micropropagated plants cultured *in vitro* for 1 or 12 months (from Leifert *et al.*, 1989a)

Plant species	1 month (explants)		>12 months (shoots/calli)	
	No. of strains	Identification	No. of strains	Identification
Astilbe	–		1	*Bacillus subtilis*
			1	*Acinetobacter calcoaceticus*
			1	*Micrococcus kristinae**
Arunchus	–		4	*Bacillus pumilus*
Choisya	3	*Agrobacterium radiobacter*	20	*Staphylococcus saprophyticus*
			2	*Staphylococcus epidermidis*
			1	*Enterobacter agglomerans**
				Erwinia sp.
Cotinus	–	–	5	*Bacillus subtilis*
			3	*Bacillus pumilus*
Delphinium	2	*Pseudomonas fluorescens*	20	*Pseudomonas maltophila*
	3	*Klebsiella oxytoca*	6	*Pseudomonas paucimobilis*
	1	*Acinetobacter calcoaceticus*	2	*Lactobacillus acidophilus**
	1	*Bacillus subtilis*	2	*Staphylococcus epidermidis**
			2	*Staphylococcus warneri*
Gerbera	4	*Pseudomonas fluorescens/putida*		
	2	*Acinetobacter calcoaceticus*		
Hemerocallis	2	*Pseudomonas paucimobilis*	24	*Lactobacillus plantarum*
	1	*Enterobacter agglomerans**	20	*Pseudomonas paucimobilis*
		Erwinia sp.	11	*Staphylococcus epidermidis*
			6	*Enterobacter cloacae*
			4	*Micrococcus kristinae**
			1	*Bacillus subtilis*
Hosta	1	*Enterobacter agglomerans**	6	*Enterobacter agglomerans**
		Erwinia sp.		*Erwinia* sp.
	2	*Pseudomonas fluorescens*	4	*Micrococcus kristinae**
	1	*Pseudomonas paucimobilis*	6	*Staphylococcus epidermidis*
			2	*Staphylococcus warneri*
Iris	7	*Erwinia carotovora*	–	–
	1	*Alcaligenes denitrificans*		
	1	*Serratia plymuthica*		
	1	*Pseudomonas fluorescens*		
Primula	–	–	6	*Bacillus circulans*
Thalictrum	–	–	3	*Bacillus subtilis*
			2	*Bacillus pumilis*
Total	33		165	

* Natural habitats are animal skin and fur.

While subliminal endophytic fungi including yeasts are, in our experience, infrequent in tissue culture, they may also be detected by culture indexing on appropriate media.

Specific tests for contaminants

Endophytic contaminants of plants and, by extrapolation, of tissue cultures include pathogenic organisms both intercellular and intracellular in distribution. Aside from the quality control aspects, phytosanitary legislation or customer requirements may demand that cultures be specifically screened for known pathogens of the crop. In the case of cultivable pathogens isolation, rather than investigation *in situ*, may facilitate diagnosis. For the detection of non-cultivable organisms *in situ* fluorescent antibody staining of tissues may be employed (Clark, this volume) but, more commonly, fluorescent antibody is used with tissue homogenates (Reuther & Sonneborn, 1983). In isolating contaminants from plants it is frequently found that more than one isolate is present (Cassells *et al.*, 1988). While it may be an exaggeration to describe this phenomenon in terms of an endogenous microflora of variable species composition, the risks associated with masking of isolates in such isolations should be appreciated.

Specific tests will be discussed here under the following broad headings: virus and virus-like agents; bacteria and bacteria-like organisms; fungi.

Viruses and virus-like agents

Historically viruses have been detected by symptom expression and diagnosed by inoculation of indicator species (Matthews, 1991). Latency is, however, a feature of virus expression *in vivo* and appears to be the norm *in vitro*. Thus, while being aware of the risk of viral contamination, pragmatically the micropropagator can only test for the presence of known viruses of the crop, for which reference can be made to *The Commonwealth Mycological Institute/Association of Applied Biologists Descriptions of Plant Viruses*.

Modern methods of rapid virus detection (Hill, 1984; Jones & Torrance, 1986) depend largely on serological tests, especially enzyme-linked immunosorbent assay (ELISA) (Clark and Adams, 1977; Torrance, this volume). ELISA kits are available for major viruses from a number of commercial suppliers (Gugerli, this volume). Alternatively, in-house antiserum may be produced (Noordham, 1973; Torrance, this volume) and antiserum of low titre can be upgraded by the use of commercial signal-enhancer kits such as are available from Amersham.

Electron microscopy may also have an application in virus screening; of particular use is immunosorbent electron microscopy (ISEM; Derrick, 1973; Roberts, 1986). However, the capital costs are high and consequently only larger micropropagation companies are likely to exploit this methodology. In carrying out virus testing the virus titre in the tissues used

is critical but this may not limit the application of ELISA to virus detection in tissue cultures, as has been verified by work on potato (Gallenberg & Jones, 1985). ELISA gives statistically positive or negative results but sensitive confirmatory tests, e.g. inoculation of indicator plants, are recommended.

DNA probes offer a sensitive alternative to serological virus detection methods (Baulcombe *et al.*, 1984; Salazar & Querci; Jones, this volume) but legal restrictions on the use of radio-labelled material restrict the use of this approach in micro-propagation companies. The use of isotopes can be circumvented by the employment of biotin-labelled probes (Leary *et al.*, 1984; Salazar & Querci; Vivian, this volume) and continuing development of non-radioactive probes is likely to make the use of probes more widely applicable.

Viroids can be detected in nuclease-treated extracts on acrylamide gels or by application of DNA probes as mentioned above (Owens & Diener, 1981; Salazar & Querci, this volume).

Bacteria and bacteria-like organisms   A wide range of microscopic, cultural, biochemical and pathogenicity tests have been used to diagnose plant pathogenic and plant-associated bacteria (Schaad, 1980; Lelliott & Stead, 1987; Stead, this volume). Recently biochemical test kits have become available from commercial sources (Stead, this volume). These can be analysed by numerical charts or computer but they do have drawbacks (Collins & Lyne, 1985) (Table 10.2). From the plant pathologist's point of view the most important of these

Table 10.2 Tests used in the identification of micro-organisms: comparative evaluation of commercial kits and individual tests (after Collins & Lyne, 1985)

The tests combined in commercial kits may be primarily constructed for the identification of bacteria from food, medical or veterinary sources rather than plants

Conventional paper strip and disc tests can be used to carry out tests specified as appropriate by plant microbiologists. Kits usually consist of 10–20 tests dedicated as above

Users evaluate their own tests whereas kits are assessed by reference to numerical charts or are interpreted by computer

If identification is by serological tests, single or few tests may be required. If identification is by biochemical tests, the more tests used the more reliable the results. In the latter circumstances, especially in the case of exotic organisms, kits may be the method of choice

Environmental and personal risk factors may vary between the methods

Conventional paper strip and disc methods are usually based on the identification methods of Cowan (1974) whereas kit methods are based on those of Edwards and Ewing (1972)

Some methods are more expensive per identification than others drawbacks is that the kits have been developed primarily for the diagnosis of bacteria from the fields of veterinary, medical and food science. Collins and Lyne (1985) argue that conventional paper strip and disc methods offer the advantage that the tests can be narrowed down to those appropriate to the most likely organisms.

As in viral diagnosis, ELISA and DNA probes have been developed for bacterial detection (e.g. Cassells *et al.*, 1988; Stead; Vivian, this volume). An alternative to in-house screening is to contract out this function to diagnostic laboratories (Cassells, 1991) which may use a combination of approaches based on those outlined above and including such modern methods as fatty acid profiling (Stead, 1988; Stead, this volume).

As mentioned above, fastidious prokaryotes, including mycoplasmas and spiroplasmas, are difficult to culture index. Such intracellular organisms may be detected by *in situ* staining with DNA specific fluorochromes, e.g. 4',6-diamidino 2-phenylindole (DAPI; Petzold & Marwitz, 1980; Clark, this volume).

Fungi   Fungal contaminants, including yeasts, are generally tolerant of plant tissue culture media; consequently they are expressed and can be detected visually. Fungal diagnosis is facilitated by the use of spore-inducing media and conditions and by reference to standard tests (e.g. Barnett & Hunter, 1972). Yeasts may be identified as above or by use of commercial test kits.

An integrated screening strategy

Preparatory stages   The objective of micropropagation is to produce healthy, true-to-type plants with acceptable habits at a competitive price (Cassells, 1991). To achieve this goal, quality control is of critical importance. The requirement of an adequate genetic base for production (Johansen *et al.*, 1984) determines in part the scope of the phytopathological services required in production as does the scale of production.

Recognition of the problems associated with the production of contaminant-free cultures ('axenic' cultures) caused Debergh and Maene (1981) to introduce a stage 0 into the micropropagation procedure. In stage 0 donor plants are prepared by culture under conditions which tend to reduce bacterial contamination. This approach may be extended, for example as in the case of pelargonium donor plants, by treatment with antibiotics (Cassells *et al.*, 1988).

In the author's view it can be beneficial to screen the prospective donor plant in stage 0 to determine its contamination load and spectrum. Heavily contaminated plants or plants contaminated with specific hazardous micro-organisms may in this way be rejected as explant donors. The work load involved in this exercise should not be underestimated as reference to a crop disease compendium such as Hooker (1981) will verify.

Following screening of potential donors, and acknowledging that an adequate genetic base for the crop will require examination of a significant number of individual plants (see above), the preferred strategy for the introduction of plants into stage 1 is via meristem tip culture. As has been discussed previously, meristem tip culture *per se* may eliminate many, or possibly all, contaminants present in the meristem donor plant (Cassells, 1986). The proviso here is that the minimum size of meristem tip compatible with establishment is excised and cultured.

Establishment of axenic cultures
Following the preparatory stage, individual cultures derived from single meristems should be subjected to rigorous screening. To provide adequate material for this purpose, it may be necessary to undertake limited multiplication of the meristems and grow the material to stage 3.

The tests to be applied should be of two types: (i) culture indexing to detect the presence of pathogenic and promiscuous cultivable micro-organisms; and (ii) specific screening for known pathogens of the crop.

Regarding specific testing, a particular risk is that the test used may be narrowly specific. Variability in prokaryotes and viruses underlies this problem. In the case of cultivable micro-organisms detection by culture indexing *per se* would indicate that the cultures should be destroyed; this confirmation is not possible with non-cultivable micro-organisms.

Production monitoring
Given that putative axenic cultures have been established, the multiplication phase (stage 2) can proceed.

Stage 2 and subsequent stages must also be monitored to reconfirm freedom from contaminants introduced in the donor tissue and also to ensure that the axenic status is not reduced by laboratory contamination. Such contamination and the spectrum of micro-organisms involved may reflect contamination originating from the person of the micropropagator, the laboratory environment, or equipment failure (Table 10.1). As has been stressed previously, non-random patterns of contamination may occur and those sampling production should bear this in mind. Five basic patterns of contamination may be encountered singly or in combination. These are as follows.

Contamination derived from stage 1: If a percentage of the stage 1 cultures are contaminated a corresponding percentage of the first and subsequent stage 2 cultures will also be contaminated. These contaminated cultures may occur in a single linear block or in smaller dispersed linear blocks depending upon work practice.

Contamination arising from poor aseptic technique: The output of individual workers may be contaminated randomly or, more commonly, downstream as a result of a deviation from strict sterile procedure. This may result in blocks of contaminated cultures, frequently towards the end of the production shift.

Contamination by airborne inoculum: Contamination of cultures in the laminar flow cabinet or in the growth rooms may occur, usually at random but also in blocks where threshold contaminant inoculum levels have been exceeded.

Contamination by media- and equipment-borne inoculum: Contamination by other than poor aseptic technique may occur where heat-resistant spores or faulty autoclaving procedures allow contamination of media and instruments. This may result in blocks of contaminated cultures.

Contamination by insect vectors: Contamination by mites may occur in the growth rooms giving rise to foci of contamination. In this case the pattern may be two- or three-dimensional.

Sampling should reflect the distribution patterns outlined above.

It is advisable to carry out random sampling at an appropriate frequency (Gilligan, 1987). To avoid serious underestimation of contamination where contaminated cultures are detected, production upstream and downstream of these points should be monitored to determine whether the points represent contaminated individual containers or blocks. Fortunately, as has been discussed above, most contaminated blocks are linear in production. The exception may be mite or other vector contamination where the contaminated cultures may be visually detected and the focus delimited.

The sampling method suggested here can be facilitated by appropriate coding of cultures. This may be achieved by the use of colour-coded price labelling systems or, more efficiently, by computer read barcoding systems (Fig. 10.1).

Progeny screening Genetic aspects of quality control (Cassells, 1991) require that production be monitored for ease of establishment, habit and trueness-to-type. This sample of production whose growth has been continued should also be monitored for disease symptoms, and culture indexed and screened for specific micro-organisms

Fig. 10.1 Labelling of cultures: (a) 'price labelling'; (b) barcoding; (c) a barcoding work station with printer and decoder/microcomputer.

(as discussed above). Also, progeny testing in this way may facilitate detection of contaminants, such as viruses, present latently in cultures in low titre. Table 10.3 outlines the proposed screening strategy.

Trends in contaminant screening in micropropagation

As the micropropagation industry matures, quality control and assurance standards are likely to become more rigorous (Cassells, 1991). Large-scale specialist producers are likely to exploit parallel developments in diagnostic methodology, and increasingly suppliers of diagnostics are catering for the plant

Table 10.3 Screening for pathogens and contaminating organisms

Stage	Action
0	Visually examine potential donor plants for disease symptoms, screen for cultivable bacteria, use streak plating to separate isolates and serial dilution to determine contamination. Use specific tests for known pathogens of the crop. Reject heavily contaminated or diseased individuals if practical
1	Visually examine cultures for contamination, including 'halo' formation. Reject all contaminated cultures. Screen remaining cultures for cultivable organisms and carry out specific tests for known pathogens of the crop. Reject all infected cultures (treat donor plants and repeat meristem culture; see Cassells, 1986; Cassells *et al.*, 1988)
2	Monitor production: sample production, examine visually and carry out tests for cultivable organisms. If contamination is detected screen upstream and downstream production and reject as appropriate
3	As for Stage 2
4	Established progeny should also be sampled and monitored principally for known diseases of the crop which may have been below the level of detection *in vitro*

Note that in carrying out the quality control steps for habit and genetic stability, screening for known pathogens of the crop should also be carried out.

Table 10.4 Health certification of *in vitro* cultures

Category	Test result*
Class 1	Free of all cultivable micro-organisms and of all major diseases of the crop
Class 2	Free of all cultivable micro-organisms and of specified diseases of the crop
Class 3	Free of cultivable bacteria
Class 4	Unscreened

* Tests must be based on adequate sampling of production. Screening for cultivable micro-organisms as in Lelliott and Stead (1987); specific tests to be based on ELISA or similar accepted tests.

sector. However, for micropropagators, regardless of scale, culture indexing is likely to be the mainstay of their phytopathological services.

In addition to the requirement for the micropropagator to monitor production for contaminants, customers and phytosanitary authorities may increasingly come under pressure to verify the micropropagators' claims regarding health status on the *caveat emptor* principle. To facilitate trade in *in vitro* microplants and cultures it is recommended that the production be certified into categories reflecting the disease-screening strategy employed (Table 10.4). While most laboratories would be capable of carrying out disease indexing, more specialized testing might be more appropriately carried out by specialized laboratories under contract.

It is further suggested that a requirement exists for phage and bacteriocin typing methods (Hawker & Linton, 1971) for plant pathogenic bacteria to allow more precise determination of the origin of contamination in the production chain from donor plant through treated mother cultures to progeny plants. This latter technology would have application where litigation over the health status of material arises.

Acknowledgements

I am most grateful to Professor W.M. Waites and his co-workers for preprints of papers in press and for permission to reproduce the data in Table 10.1.

References

Barnett H.L. & Hunter B.B. (1972) *Illustrated Genera of Imperfect Fungi.* Burgess, Minneapolis.

Bastiaens L. (1983) Endogenous bacteria in plants and their implications in tissue culture — a review. *Mededelingen van de Faculteit Landbouwwetenschappen Rijksuniversiteit Gent* **48**, 1–11.

Baulcombe D., Flavell R.B., Boulton R.E. & Jellis G.J. (1984) The sensitivity and specificity of a rapid nucleic acid hybridization method for the detection of potato virus X in crude sap samples. *Plant Pathology* **33**, 361–70.

Bove J.M. (1988). Plant mollicutes: phloem-restricted agents and surface contaminants. *Acta Horticulturae* **225**, 215–22.

Campbell R. (1985) *Plant Microbiology.* Arnold, London.

Cassells A.C. (1986) Production of healthy plants. In Alderson P.G. & Dullforce W.M. (eds) *Micropropagation in Horticulture: Practice and Commercial Problems.* Institute of Horticulture, London.

Cassells A.C. (1991) Setting up a commercial micropropagation laboratory. In Bajaj Y.P.S. (ed.) *Biotechnology in Agriculture and Forestry*, pp. 53–70. Springer-Verlag, Berlin.

Cassells A.C., Harmey M.A., Carney B.F., McCarthy E. & McHugh A. (1988) Problems posed by cultivable bacterial endophytes in the establishment of axenic cultures of *Pelargonium* × *domesticum*: the use of *Xanthomonas pelargonii*-specific ELISA, DNA probes and culture indexing in the screening of antibiotic treated and untreated donor plants. *Acta Horticulturae* **225**, 153–62.

Clark M.F. & Adams A.N. (1977) Characteristics of the microplate method of enzyme-linked immunosorbent assay for the detection of plant viruses. *Journal of General Virology* **34**, 475–583.

Collins C.H. & Lyne P.M. (1985) *Microbiological Methods*, 5th edn. Butterworths, London.

Cowan S.T. (1974) *Cowan and Steel's Manual for the Identification of Medical Bacteria*, 2nd edn. Cambridge University Press, Cambridge.

Debergh P.C. & Maene L.J. (1981) A scheme for commercial propagation of ornamental plants by tissue culture. *Scientia Horticulturae* **14**, 335–45.

Debergh P.C. & Vanderschaeghe A.M. (1988) Some symptoms indicating the presence of bacterial contaminants in plant tissue cultures. *Acta Horticulturae* **225**, 77–82.

Derrick K.S. (1973) Quantitative assay for plant viruses using serologically specific electron microscopy. *Virology* **56**, 652–3.

Dimock A.W. (1962) Obtaining pathogen-free stock by cultured cutting techniques. *Phytopathology* **52**, 1239–41.

Edwards P.R. & Ewing W.H. (1972) *Identification of Enterobacteriaceae*, 3rd edn. Burgen, Minneapolis.

Gallenberg D.J. & Jones E.D. (1985) Detection of potato viruses X and S in tissue culture plants. *American Potato Journal* **62**, 118–19.

Gilligan C.A. (1987) Analysis of spatial patterns of soil-borne pathogens. In Kranz J. & Rotem J. (eds) *Experimental Techniques in Plant Disease Epidemiology*, pp. 85–98. Springer-Verlag, Heidelberg.

Hawker L.E. & Linton A.H. (eds) (1971) *Microorganisms: Function, Form and Environment*, 2nd edn. Edward Arnold, London.

Hayward A.C. (1974) Latent infections by bacteria. *Annual Reviews of Phytopathology* **12**, 87–97.

Hill S.A. (1984) *Methods in Plant Virology*, Blackwell Scientific Publications, Oxford.

Hooker, W.J. (ed.) (1981) *Compendium of Potato Diseases*. American Phytopathological Society, St. Paul, Minnesota.

Johansen D.G., Knutson K.W., Slack S.R. & Jones E.D. (1984) Recommendations for standards and guidelines for acceptance of tissue cultured potatoes into certification programmes. *American Potato Journal* **61**, 368–70.

Jones R.A.C. & Torrance L. (eds) (1986) *Developments and Applications in Virus Testing*. Association of Applied Biologists, Wellesbourne.

Leary J.J., Brigati D.J. & Ward D.E. (1984) Rapid and sensitive colorimetric method for visualizing non-labelled DNA probes hybridized to DNA or RNA immobilized on nitro-cellulose bioblots. *Proceedings of the National Academy of Science, USA* **80**, 4045–9.

Leifert C., Waites W.M. & Nicholas J.R. (1989 a) Bacterial contaminants of micropropagated plant cultures. *Journal of Applied Bacteriology* **67**, 353–61.

Leifert C., Waites W.M., Camotta H. & Nicholas J.R. (1989 b) *Lactobacillus plantarum*: a deleterious contaminant of plant tissue cultures. *Journal of Applied Bacteriology* **67**, 363–70.

Lelliott R.A. & Stead D.E. (1987) *Methods for the Diagnosis of Bacterial Diseases of Plants*. Blackwell Scientific Publications, Oxford.

Long R.D., Curtin T.F. & Cassells A.C. (1988) An investigation of the effects of bacterial contaminants on potato nodal cultures. *Acta Horticulturae* **225**, 83–92.

Matthews R.E.P. (1991) *Plant Virology*. 3rd edn, Academic Press, New York.

Menard D., Coumans M. & Gaspar Th. (1985) Micropropagation du *Pelargonium* à partir de méristemes. *Medelingen van de Faculteit Landbouwwetenschappen Rijksuniversiteit Gent* **50**, 327–31.

Murant A.F. & Harrison B.D. (eds) *Descriptions of Plant Viruses*, CMI/AAB 1970 onwards. Commonwealth Mycological Institute and Association of Applied Biologists, Wellesbourne.

Murashige T. (1974) Plant propagation through tissue culture. *Annual Review of Plant Physiology* **25**, 135–66.

Noordham D. (1973) *Identification of Plant Viruses: Methods and Experiments*. Centre for Agricultural Publishing and Documentation, Wageningen.

Owens R.A. & Diener T.O. (1981) Sensitive and rapid diagnosis of potato spindle tuber viroid disease by nucleic acid hybridization. *Science* **213**, 670–2.

Petzold H. & Marwitz R. (1980) Ein verbesserter fluoreszenzmikroskopischer

Nachweis für mycoplasmaähnliche Organismen in Pflanzengewebe. *Phyto-pathologische Zeitschrift* **97**, 327–31.

Reuther G. (1985) Principles and application of the micropropagation of ornamental plants. In Schafer-Menuhr A. (ed.) *In Vitro Techniques: Propagation and Longterm Storage*, pp 1–14. Martinus Nijhoff, Dordrecht.

Reuther G. & Sonneborn H.H. (1983) Untersuchungen zur fruhdiagnose von *Xanthomonas* infektion bei Pelargonien in gewebekulture. *Gartenbau Gartenwissenschaften* **33**, 854–7.

Roberts I.M. (1986) Practical aspects of handling, preparing and staining samples containing plant virus particles for electron microscopy. In Jones R.A.C. & Torrance L. (eds) *Developments and Applications in Virus Testing*, pp. 213–43. Association of Applied Biologists, Wellesbourne.

Roberts S.J. & Ambler D.J. (1988) Progress with identification methods for a fastidious xylem-limited bacterium which causes Sumatra disease of cloves in Indonesia. *Acta Horticulturae* **225**, 103–8.

Schaad N.W. (1980) *Laboratory Guide for Identification of Plant Pathogenic Bacteria*. American Phytopathological Society, St. Paul, Minnesota.

Stead D.E. (1988) Identification of bacteria by computer-assisted fatty acid profiling. *Acta Horticulturae* **225**, 39–46.

11 Application of rapid techniques to seed health testing — prospects and potential

S. BALL & J. REEVES

Introduction

Epidemiology Seeds can be effective disease dispersal agents. Many disease organisms can be harboured in or on seed and may remain viable for years under suitable storage conditions. There can be a high disease transmission rate from seed to crop; even if not all infected seeds result in diseased plants, a high proportion of infected seed can still produce sufficient disease foci in the field to initiate serious epidemics (Hewett, 1983).

As countries diversify into more exotic crops, and trade barriers are removed, particularly in Europe, the opportunities for introduction of pathogens to new areas via imported seed are increasing (Bos, 1977; Hampton, 1983). Commercial seed production is big business and within national as well as international seed trading there is potential for transfer of seed-borne pathogens.

Legislative control Entry restrictions on proscribed organisms in seed and removal of infected seed stocks in early generations in the multiplication process can prevent waste of resources and crop losses later on. Consequently quarantine and certification schemes exist in most countries to ensure flow of good seed within and between countries. The Food and Agriculture Organization (FAO) instigated the Quality Declared Scheme which sets minimum standards in its programme for improvement of food supplies, particularly in the developing world (Kelly, 1989).

The International Seed Testing Association (ISTA) is an intergovernmental organization which develops and publishes standard procedures for the sampling and testing of seeds traded internationally, provides strict rules for seed purity and germination and has also published guidelines for testing some of the most important seed-borne diseases (Richardson, 1979; *ISTA Handbook on Seed Health Testing*). However, the scope of seed pathology and range of test methods are too great to impose universal rules.

The presence or absence of seed-borne diseases can determine whether seed can be imported or exported and to what certification category it may be assigned (Hewett, 1966). Premiums may be paid for certified seed whereas failure to comply with the specified standards can mean downgrading or conversion to other less profitable uses such as feed. The current range of diseases and standards covered by UK legislation are shown in Table 11.1.

Seed health testing

Seed usually becomes infested during growth of the crop in the field, but there are also opportunities for contamination during the post-harvest processing of seeds for further multiplication in the series of cleaning, sorting and storage procedures. Combine harvesters deliver the material from the field in trailers to farm buildings where the seeds are passed through seed-cleaning machines to remove most of the non-seed debris and broken seed. They are then dried if necessary, either on the floor or in bins or silos. At this stage, before further cleaning or movement, a sample may be taken and tested for germination and certain disease organisms. Most of the disease testing is done by specialist laboratories within the Official Seed Testing Stations (OSTS). After transport to processors and merchants, and the final cleaning, treating, bagging and labelling, the seed is divided into specified lots. Official samples are drawn from the final seed lots and tested for compliance with the standards prescribed in the UK Seeds Regulations (1985) under the Plant Varieties and Seeds Act (1964) before marketing. Seed lots may be part of a large single crop or may originate from several crops.

For pathogens which can be carried with seed, such as some smuts, bunts, downy mildews and bacteria, there are risks of cross-contamination during processing, and rapid confirmation of their presence together with stringent hygiene measures are therefore essential. Testing can also be used to determine whether or not seed treatment is required, and what would be the most appropriate and economical treatment (Hewett & Griffiths, 1986).

Other areas where there is a need for testing for the presence or absence of disease organisms include research work and surveys. However, it is important to recognize that a diagnostic test that works well on a small scale in a research laboratory needs to meet more demanding criteria when applied commercially. In many crops seed is sown and harvested in the same year, for example, winter cereals and winter pulses, which means a period of intense activity to process and certify large numbers of seed lots as quickly as possible to meet sowing deadlines. Routine test methods must therefore be geared to deal with high throughputs and rapid turnover in addition to meeting standards of specificity and sensitivity appropriate to the epidemiology of the diseases being tested for.

In some countries routine methods for seed testing have to be performed in laboratories where there is little equipment and by staff with limited training. Simple, relatively fool-proof test methods are therefore required for routine applications, preferably with referral to a specialist to monitor the quality of results and to confirm doubtful cases.

Until the mid 1980s most seed testing methods for fungi were adaptations of standard microbiological techniques such as agar plate and moist chamber (blotter) cultures with direct observation of culture morphology and microscopic examination of reproductive structures after a period of incubation (Neergard, 1977; Agarwal & Sinclair, 1987). Tests for seed-borne bacteria consisted of isolation, biochemical tests (Stead, this volume) and host inoculation. Tests for seed-borne viruses also consisted of growing-on seed (Rohloff, 1967) or inoculation of indicator plants with extracts of ground seed (Marrou & Messiaen, 1967; Kimble *et al.*, 1975). More recently enzyme-linked immunosorbent assays (ELISAs) and other serological techniques (Torrance; Stead; Dewey, this volume) have been adopted for some seed-borne diseases, e.g. lettuce mosaic potyvirus (LMV; van Wuurde & Maat, 1983), pea seed-borne mosaic potyvirus (Maury *et al.*, 1987) and others (Lange & Heide, 1986; Sheppard *et al.*, 1986).

Some of these methods will be discussed in more detail to demonstrate some of the problems which the newer techniques may help to resolve.

Detecting phytopathogenic micro-organisms in seed

Growing-on test This is one of the least complex means of testing for any seed-borne pathogen. It involves the growing of a sample of seed and assessing the disease incidence in the seedlings. Provided that no contamination occurs after sowing, an estimate of the level of disease present in the seed lot can be obtained, although it may not distinguish between disease carried on the seed coat from that in the embryo. The growing-on test requires few technical resources and is easily done but it is not very sensitive. Large numbers of seeds (up to 10 000) may need to be sown to test for low levels of disease, especially with bacterial and viral diseases; this uses up large areas of glasshouse space. It is also time-consuming, relies on the viability of the seed under test and on the unambiguous expression of disease symptoms. Growing-on tests are generally not appropriate where large numbers of seed samples need to be tested rapidly.

Table 11.1 UK regulations on standards for levels of disease in seed certification
(a) Non-cereal crop standards

Seeds as per regulations	Crop	Disease	Seed category	Disease standard	Quantity of seed examined
Fodder	Field beans	*Ascochyta fabae*	Pre-basic	1 seed	1000
			Basic	2 seeds	1000
			C1	2 seeds	500
			C2	1%	
	Peas	*Pseudomonas syringae* pv. *pisi*	All	Nil	1 kg
Oil and fibre	Linseed	*Botrytis* spp. *Alternaria linicola*	All	5%	150 g
		Phoma exigua var. *linicola Colletotrichum lini Fusarium* spp.	All	5% (1% in flax)	
	Turnip rape	*Sclerotinia sclerotiorum*	All	5 sclerotia	70 g
	Swede rape	*Sclerotinia sclerotiorum*	All	10 sclerotia	100 g
	White mustard	*Sclerotinia sclerotiorum*	All	5 sclerotia	200 g
	Sunflower	*Botrytis* spp.	All	5%	1 kg
Vegetables	Brassicas	*Phoma lingam*	Pre-basic & basic	Nil *	1000
	Red beet	*Phoma betae*	Pre-basic & basic	Nil *	200
	French beans	*Colletotrichum lindemuthianum*	Pre-basic & basic	Nil	600
		Pseudomonas syringae pv. *phaseolicola*	Pre-basic & basic	Nil	5000

Crop	Pathogen	Seed category	Standard	Number
Peas	*Ascochyta* spp.	Pre-basic, basic & certified	Nil *	200
Broad beans	*Ascochyta fabae*	Pre-basic & basic	Nil *	600
Lettuce	Lettuce mosaic potyvirus	Pre-basic, basic & certified	Nil	5000
Celery	*Septoria apiicola*	Pre-basic, basic & certified	Nil *	400
	Phoma apiicola	Pre-basic, basic & certified	Nil *	400

* Can be retrieved by treatment.

(b) Cereal crop, standards

Disease	Seed category	Minimum standard (EEC directives)	Higher voluntary standard (HVS)
Ergot	Pre-basic & basic	1 piece in 500 g	Nil in 1 kg
	Certified	3 pieces in 500 g	1 in 1 kg
Loose smut (*Ustilago nuda*)	Pre-basic & basic	0.5%	0.1% in 1000
	Certified	0.5%	0.2% in 1000

Direct plating   Seeds may be surface sterilized and plated directly on to agar media appropriate for the pathogens to grow. This simple method is commonly used to detect fungal pathogens such as *Ascochyta* spp. in peas and beans where it is relatively easy to identify the pathogen visually. It is also used for bacteria where growth on a semi-selective medium is taken as a presumptive identification which may be confirmed by further testing (Schaad & Kendrick, 1975).

Host plant inoculation   This method does not always require the isolation of a presumed pathogen on an agar medium. A suitable indicator host plant can be sprayed or injected with a crude seed extract containing an abrasive compound to facilitate entry of the disease, for example when testing for LMV (Marrou & Messiaen, 1967). Host inoculation can also be used in the final stages of testing to confirm the identity, pathogenicity or viability of organisms isolated on agar plates, for example *Septoria apiicola* in celery seeds (Maude, 1963; Hewett, 1968), and bacterial blight of pea (Malik *et al.*, 1987).

Immunodetection   Pathogenic organisms can be detected directly in seed extracts by immunofluorescence (IF) microscopy and ELISA; these techniques are widely used in routine testing (van Wuurde *et al.*, 1983). They generally work well when used in testing for viruses but are less successful with bacteria and fungi. This is because of the failure to achieve absolute specificity with polyclonal antisera and cross-reactions with other, non-target organisms. Specificity can be improved by the use of monoclonal antibodies (MAbs) (Stead; Dewey, this volume), but they are expensive to produce and cross-reactions may still occur owing to the ubiquity of the epitope to which the MAb was raised as, for example, within the generic groups *Pseudomonas* and *Pythium*. Other techniques such as the Ouchterlony double-diffusion in gel test can be used with seed or tissue extracts (De Boer, 1989) but these also suffer from problems with specificity.

Detection of phytopathogenic bacteria

A range of the techniques used for detecting and identifying bacteria are described by Stead (this volume). Testing for seed-borne bacterial pathogens can involve three stages: (i) extraction from the seed; (ii) isolation on agar plates; and (iii) identification. In practice the last two steps often overlap.

Extraction   Clearly the extraction procedure is very important because, if inappropriate, it would prejudice the sensitivity of the test. In a recent review, Roth (1989) outlined the major factors which need consideration before adopting a particular method. Most techniques involve soaking a quantity of seed in a defined extraction medium under conditions of temperature, time, volume and aeration which optimize maximal recovery of the target organism with minimal stimulation of the background microflora. Surface sterilization of the seed before extraction can reduce the latter but must be used with circumspection as many pathogens may only be present on the seed surface (Taylor *et al.*, 1979). Where the pathogen is harboured within the seed, some comminution of the seed prior to extraction may be appropriate. The degree and amount of comminution may also have an effect on the recovery of the target organism (J. Reeves, unpublished results).

Isolation   The liquid extract can be plated out (usually at different dilutions) onto agar media, the choice of which depends on the target organism. In addition to general plating media, semi-selective media can be used for this work (Schaad, 1988).

Identification   Many biochemical tests exist for the identification of bacteria (Stead, this volume), very often based on the ability to use a carbon source. Different attributes may be examined depending on the target organism: a good example is the LOPAT scheme for the presumptive identification of green-fluorescent pseudomonads (Lelliott *et al.*, 1966). A more comprehensive scheme for *Pseudomonas* is given by Hildebrand *et al.* (1988) which lists various selective media for use in isolation techniques.

Older serological techniques such as slide or tube agglutination can be used as well as IF or ELISA, but again all suffer from the disadvantages of cross-reactivity to some degree, and further testing of serologically positive isolates may be required, usually by inoculation to host plants.

The final stage of identification often involves pathogenicity tests with inoculation and growing-on of host plants under defined conditions. Pure bacterial cultures can be introduced into the host in many ways: by spray or injection; by leaf abrasion with carborundum mixtures; by vacuum infiltration; or under high pressure. It is important that a test gives results which are easily interpreted, and preferably symptoms which resemble those produced by the pathogen under field conditions. Pathovars of *Pseudomonas syringae* are distinguished by their reaction on the appropriate host species, and races by their reactions on different cultivars of the host species (Taylor *et al.*, 1989). Clearly the choice of species or cultivar can be crucial to the success of pathogenicity tests. Should equivocal results occur, then other data such as serological reactions should also be used before reaching any conclusions.

Bacterial blight of pea — a case history

Bacterial blight of pea (*Pseudomonas syringae* pv. *pisi*) is primarily seed-borne. It was first discovered in the UK in 1985 in the pea cultivar Belinda (Stead & Pemberton, 1987). To minimize its spread the Ministry of Agriculture, Fisheries and Food (MAFF) instituted a policy in 1986 of testing seed samples in early generations of seed multiplication, and this became a legal requirement in 1987. To prevent spread of the disease, samples are taken and tested before movement of the crop from the farm. Consequently hundreds of 1 kg samples for certification (and large numbers for advisory purposes outside the scheme) need to be tested rapidly since the seed must be processed and sold as soon as possible after harvest. Details of the test methods are given here (Fig. 11.1) since they demonstrate the multi-stage nature of testing as well as illustrating the sorts of problems and approaches encountered with other diseases mentioned earlier. The test method for phytopathogenic pseudo-monads was developed in the UK (Taylor, 1970, 1984) and although effective, it takes about 2 weeks to complete and is labour intensive.

Bacterial blight of pea is primarily seed-borne and so the removal of infected seed lots from the multiplication cycle should control it. With 1 year of testing the frequency of infected seed lots declined from 12% to 4% (J. Reeves & S. Ball, unpublished results). However, in recent years it has increased again, not because of the inadequacy of seed lot testing *per se*, but through loopholes in the regulations which can result in some infected seed remaining in the system.

New developments in test methods

Existing test methods are multi-stage, labour-intensive and time-consuming. Very often subjective assessments may be required, calling for a high level of knowledge and experience from the personnel involved. Future improvements in the situation will most likely depend upon the increasing application and refinement of immunological and nucleic acid technologies.

Most improvements in seed-testing methods have concentrated on identifying the target organisms after isolation into pure culture. If there is no need to retain an isolate of the pathogen for further study or deposition in a culture collection, ideally the pathogen should be detected directly in the seed or after simple extraction from it with no loss of sensitivity. Even

EXTRACTION

ISOLATION

IDENTIFICATION

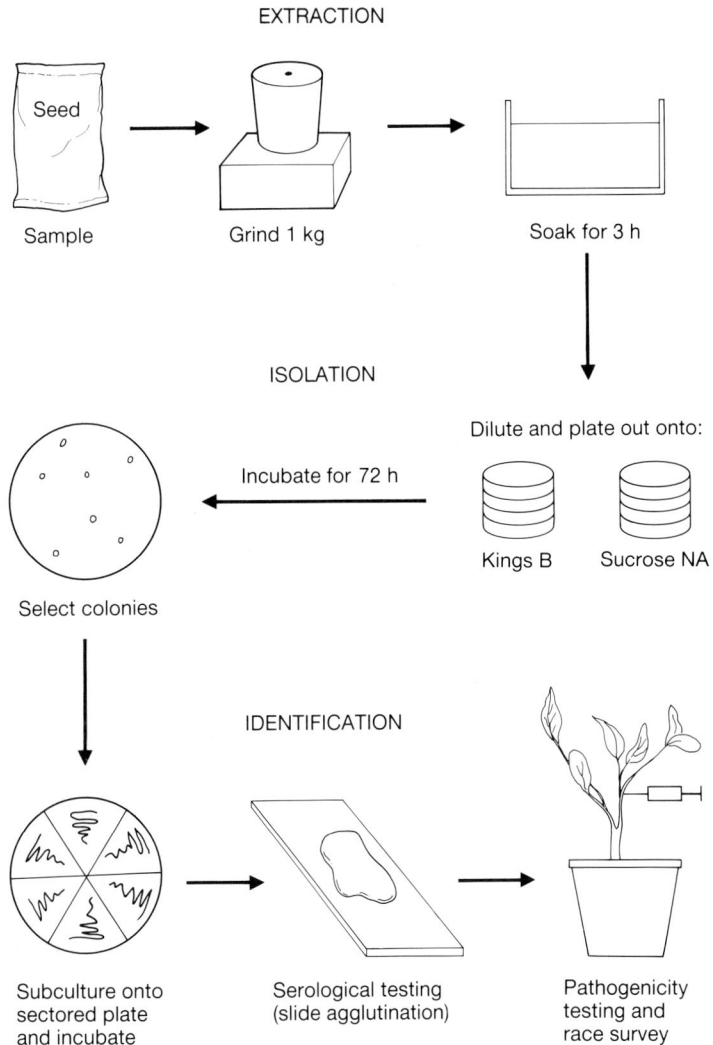

Fig. 11.1 Schematic representation of the protocol for testing pea seed for bacterial blight of pea.

if an isolate was required, direct testing would eliminate negative samples. It would also reduce labour input, test duration and costs by increasing the number of samples tested at one time, and the range of tests performed. More rapid production of results for clients would also reduce their costs.

Immunological techniques

Immunodetection and serological identification is highly effective in detecting virus infection of seed, e.g. LMV and pea early browning tobravirus (van Wuurde & Maat, 1985), squash mosaic comovirus in cucumber seeds (Nolan & Campbell, 1984) and others (Sheppard *et al.*, 1986; Maury & Khetarpal, 1989). However, the required level of specificity for anything more than a presumptive identification of bacteria and fungi has not yet been achieved and pathogenicity testing is generally required.

This situation may be changing as more experience is gained with MAbs. A MAb-based test has recently been developed for bacterial blight of pea (Candlish *et al.*, 1988) and Dewey (this volume) has shown that fungi contaminating rice grains can be easily detected and identified. Also, Mitchell (1988) has used a mixture of MAbs to detect the pathogen *Sirococcus strobilinus* in extracts from infected spruce seeds. There may therefore be room for a much wider application of MAb technology to the detection and identification of bacteria and fungi associated with seed. The ease and ubiquity of ELISA or dot-blot tests make them ideal candidates for large-scale commercial testing in kit form, but they cannot be applied until suitable MAbs, with the required sensitivity and specificity, have been obtained and widely tested.

Nucleic acid probes   Recent developments in molecular biology have provided other novel means of identifying various organisms, in particular DNA probes and the polymerase chain reaction (PCR) (Vivian; Salazar & Querci; Coddington & Gould, this volume). Their use in clinical microbiology has been comprehensively reviewed by Tenover (1988) who listed 16 commercially available kits based on nucleic acid probes for the identification of human pathogens; viruses, mycobacteria and bacteria. The problems associated with the detection of seed-borne organisms are not dissimilar to those encountered in a clinical microbiology laboratory where there is also a need for accuracy, speed and economy. In examining the cost effectiveness of DNA probes, Tenover (1988) identified radioactive labelling as a major disadvantage. The half-lives of ^{32}P or ^{125}I reduce the shelf life of the kits, and handling them safely increases costs. There is every expectation that this problem will be overcome by non-isotopic labelling methods (Salazar & Querci; Vivian, this volume), although, at the time of writing, none yet have the sensitivity of ^{32}P-labelled probes. Estimated costs per sample tested varied from organism to organism and Tenover (1988) concluded that costs depended on local laboratory requirements. Only where large numbers of samples are to be tested is DNA-probe technology likely to show a direct cost advantage, although there will be indirect benefits in the form of speed and accuracy which need to be taken into account. The economics of the latter are not easy to assess but are of considerable importance to laboratory managers and will weigh heavily in the choice of test technique. Clearly the choice will be easier for a large laboratory testing many samples for the same organism, and Tenover (1988) recommended that smaller laboratories should send their samples to such a centre.

Although DNA probes have been applied widely and quite quickly in research programmes in plant pathology, particularly on plant viruses but also on bacteria and fungi, they have not moved into routine disease testing so quickly. Some reports have described their use in identifying particular pathogens: Schaad *et al.* (1986) reported on a probe for *Pseudomonas syringae* pv. *phaseolicola* and, more recently, they have indicated its value in routine testing (Schaad *et al.*, 1989). However, they too have identified isotopic labelling as costly and disadvantageous. Probes have also been developed for taxonomic studies on *P.fluorescens* (Festl *et al.*, 1986), and to differentiate between *P.syringae* pv. *tomato* and *P.syringae* pv. *syringae* (Denny, 1988). The potential uses of probes are considerable and cover areas of interest in seed testing such as the identification of bacteria (Vivian, this volume) and distinction between races of fungal pathogens (Coddington & Gould, this volume), but most reports on probes do not make clear whether probes would fulfil this potential in large-scale testing. Varveri *et al.* (1988), using a probe for plum pox potyvirus, reported that their molecular hybridization technique was superior to ELISA in routine use, but they gave no details of relative costs. It is claimed that a cDNA probe for detection of potato viruses X and Y in sap-hybridization tests, which have been used to supplement ELISAs, reduces costs in large-scale routine test programmes (Boulton *et al.*, 1986).

Where direct testing of micro-organisms in seed washings, sap, seed extracts and other material is possible, nucleic acid probes are likely to generate savings over existing detection methods. However, in testing for bacteria, where the organism may still have to be isolated and purified before probing, more development will be needed for maximum cost-effectiveness. For example, Schaad *et al.* (1989) reported that a probe, effective on isolated colonies, gave variable results when used in liquids containing low numbers of target cells or where the concentration of the background microflora was high.

The Official Seed Testing Station (OSTS) in Cambridge is the major UK agent for seed testing and has to test the very large numbers of samples identified by Tenover (1988) as being important for economy. A routine test being developed at OSTS and based on a DNA probe for *Pseudomonas syringae* pv. *pisi* should allow the economics and practical difficulties of using DNA probes for phytopathogenic bacteria to be evaluated.

Polymerase chain reaction   The PCR reported by Saiki *et al.* (1985) and Saiki *et al.* (1988) and described by Mullis & Faloona (1987) is a sophisticated technique involving the amplification of a specific piece of DNA from the genome of a target cell. It offers a means for detecting the presence of pathogenic organisms present in very low concentrations in infected tissue and has been already exploited successfully in this way by medical microbiologists to detect viral pathogens (Ou *et al.*, 1988; Shibata *et al.*, 1988) and, more recently, pathogenic bacteria (Hartskeerl *et al.*, 1989). The potential increase in sensitivity provided by the PCR technique has resulted in considerable interest and many workers are finding it an invaluable tool for detecting and identifying organisms, although there are as yet few reports of its use in the detection of plant pathogens. Nonetheless, Rollo *et al.* (1987) used PCR to identify *Phoma tracheiphila*, the causal agent of drying disease of lemon trees. Although PCR has great potential and may be particularly suited for the detection of plant pathogens which are often present in low numbers compared to background micro-organisms, the costs are likely to be high. The capital equipment is not very expensive but it requires costly chemicals and reagents. Nevertheless, these costs will be justified by expected increases in turnover, reductions in test time and through other savings.

Conclusions

Applications of PCR continue to proliferate. Modifications to both nucleic acid and serological techniques are being developed rapidly. Combinations of techniques may prove more effective than any individual technique. Single-domain antibodies produced with the aid of PCR (Ward *et al.*, 1989) could supplant MAbs in some instances. Clearly, further research is necessary to establish specificities and applications, particularly for transfer to the plant kingdom. However, the developments in recent years of the molecular biological techniques for detection and identification of all types of organisms have shown their potential power. For this potential to be realized in routine testing for seed-borne pathogens more development and full assessment of cost-effectiveness is required. It is likely that as methods are simplified costs will decrease and these modern techniques will become the tools of choice for many seed-testing laboratories, particularly for detection of micro-organisms not easily distinguished by the traditional methods. In seed-testing technology where both speed and accuracy are paramount, improved technology will be welcomed.

References

Agarwal V.K. & Sinclair J.B. (1987) *Principles of Seed Pathology*. CRC Press Inc., Boca Raton, Florida.

Bos L. (1977) Seedborne viruses. In Hewitt W.B. & Chiarappa L. (eds) *Plant Health Quarantine in International Transfer of Genetic Resources*, pp. 36–9. CRC Press Inc., Cleveland, Ohio.

Boulton R.E., Jellis G.J., Baulcombe D.C. & Squire A.M. (1986) The application of complementary DNA probes to routine virus detection with particular reference to potato viruses. In Jones R.A.C. & Torrance L. (eds) *Developments and Applications in Virus Testing*, pp. 41–53. Association of Applied Biologists, Wellesbourne.

Candlish A.A.G., Taylor J.D. & Cameron J. (1988) Immunological methods as applied to bacterial pea blight. *Proceedings of the 1988 Brighton Crop Protection Conference − Pests and Diseases*, pp. 787–94. British Crop Protection Council, Thornton Heath, Surrey.

De Boer S.H. (1989) Detection of *Clavibacter michiganense* subsp. *sepedonicum* in potato. In Saettler A., Schaad N.W. & Roth D.A. (eds) *Detection of Bacteria in Seed and Other Planting Material*, pp. 92–107. The American Phytopathological Society, St. Paul, Minnesota.

Denny T.P. (1988) Differentiation of *Pseudomonas syringae* pv. *tomato* from *P.s. syringae* with a DNA hybridisation probe. *Phytopathology* **78**, 1186–93.

Festl H., Ludwig W. & Schleifer K.H. (1986) DNA hybridisation probe for the *Pseudomonas fluorescens* group. *Applied and Environmental Microbiology* **52**, 1190–4.

Hampton R.O. (1983) Seedborne viruses in crop germplasm resources, disease dissemination risks and germplasm reclamation technology. *Seed Science and Technology* **11**, 535–46.

Hartskeerl R.A., De Wit M.Y.L. & Klatser P. (1989) Polymerase chain reaction for the detection of *Mycobacterium leprae*. *Journal of General Microbiology* **135**, 2357–64.

Hewett P.D. (1966) The development of seed health testing in England and Wales. *NAAS Quarterly Review*, No. 73, 1–6.

Hewett P.D. (1968) Viable *Septoria* spp. in celery seed samples. *Annals of Applied Biology* **61**, 89–98.

Hewett P.D. (1983) Epidemiology − fundamental for disease control. *Seed Science and Technology* **11**, 697–706.

Hewett P.D. & Griffiths D.C. (1986) Biology of seed treatment. In Jeffs K.A. (ed.) *Seed Treatment*, pp. 7–15 British Crop Protection Council, Thornton Heath, Surrey.

Hildebrand D.C., Schroth M.N. & Sands D.C. (1988) In Schaad N.W. (ed.) *Laboratory Guide for Identification of Plant Pathogenic Bacteria*, 2nd edn. The American Phytopathological Society, St. Paul, Minnesota.

Kelly A. (1989) *Seed Planning and Policy for Agricultural Production. The Roles of Government and Private Enterprise in Supply and Distribution*. Pinter Publishers Ltd., London.

Kimble K.A.T., Grogan R.G., Greathead A.S., Paulus A.O. & House J.K. (1975) Development, application and comparison of methods for indexing lettuce seed for mosaic virus in California. *Plant Disease Reporter* **59**, 461–6.

Lange L. & Heide M. (1986) Dot immuno-binding (DIB) for detection of virus in seed. *Canadian Journal of Plant Pathology* **8**, 373–9.

Lelliott R.A., Billing E. & Hayward A.C. (1966) A determinative scheme for the fluorescent plant pathogenic pseudomonads. *Journal of Applied Bacteriology* **29**, 470–89.

Malik A.N., Vivian A. & Taylor J.D. (1987) Isolation and characterisation of three classes of mutants in *Pseudomonas syringae* pathovar *pisi* with altered behaviour towards their host, *Pisum sativum*. *Journal of General Microbiology* **133**, 2393–9.

Marrou J. & Messiaen C.M. (1967) The *Chenopodium quinoa* test: a critical method for detecting seed transmission of lettuce mosaic virus. *Proceedings of the International Seed Testing Association* **32**, 49–57.

Maude R.B. (1963) Testing the viability of *Septoria* on celery seed. *Plant Pathology* **12**, 15–17.

Maury Y. & Khetarpal R.K. (1989) Testing seed for viruses using ELISA. In Agnihotri V.P., Singh N., Chaube H.S., Singh U.S. & Dwivedi T.S. (eds) *Perspectives in Plant Pathology*, pp. 31–49. Today and Tomorrow Printers and Publishers, New Delhi.

Maury Y., Bossenec J.M., Boudazin G., Hampton R.O., Pieterson G. & Maguire J.D. (1987) Factors influencing ELISA evaluation of transmission of pea seedborne mosaic virus in infected pea seed: seed group size and decortication. *Agronomie* **7**, 225–30.

Mitchell L.A. (1988) A sensitive dot immunoassay employing monoclonal antibodies for detection of *Sirococcus strobilinus* in Spruce seed. *Plant Disease* **72**, 664–7.

Mullis K.B. & Faloona F.A. (1987) Specific synthesis of DNA *in vitro* via a polymerase-catalysed chain reaction. *Methods in Enzymology* **155**, 335–50.

Neergard P. (1977) *Seed Pathology*. Macmillan, London.

Nolan P.A. & Campbell R.N. (1984) Squash mosaic virus detection in individual seeds and seed lots of cucurbits by enzyme-linked immunosorbent assay. *Plant Disease* **68**, 971–5.

Ou C-Y., Kwok S., Mitchell S.W. *et al.* (1988) DNA amplification for direct detection of HIV-1 in DNA of peripheral blood mononuclear cells. *Science* **239**, 295–7.

Richardson M.J. (1979) *An Annotated List of Seed-borne Diseases*. Commonwealth Mycological Institute, Phytopathological Paper, No. 23. Also published as *ISTA Seed Health Testing Handbook*, 3rd edn. International Seed Testing Association, Zurich.

Rohloff I. (1967) The controlled environment room test of lettuce seed for identification of lettuce mosaic virus. *Proceedings of the International Seed Testing Association* **32**, 59–63.

Rollo F., Salvi R., Amici A. & Anconetani A. (1987) Polymerase chain reaction fingerprints. *Nucleic Acids Research* **15**, 9094.

Roth D.A. (1989) Review of isolation and extraction methods. In Saettler A.W., Schaad N.W. & Roth D.A. (eds) *Detection of Bacteria in Seed and Other Planting Material*, pp. 3–8. The American Phytopathological Society, St. Paul, Minnesota.

Saiki R.K., Scharf S., Faloona F. *et al.* (1985) Enzymatic amplification of B-globin genomic sequences and restriction site analysis for diagnosis of sickle cell anaemia. *Science* **230**, 1350–4.

Saiki R.K., Gelfand D.H., Stoffel S. *et al.* (1988) Primer directed enzymatic amplification of DNA with a thermostable DNA polymerase. *Science* **239**, 487–91.

Schaad N.W. (1988) *Laboratory Guide for Identification of Plant Pathogenic Bacteria*, 2nd edn. The American Phytopathological Society, St. Paul, Minnesota.

Schaad N.W. & Kendrick R. (1975) A qualitative method for detecting *Xanthomonas campestris* in crucifer seed. *Phytopathology* **65**, 1034–6.

Schaad N.W., Azad H., Peet R.C. & Panopoulos W.J. (1986) Cloned phaseolotoxin gene as a hybridisation probe for identification of *Pseudomonas syringae* pv. *phaseolicola*. *Phytopathology* **76**, 846 (abs.).

Schaad N.W., Azad H., Peet R.C. & Panopoulos N.J. (1989) Identification of *Pseudomonas syringae* pv. *phaseolicola* by a DNA hybridisation probe. *Phytopathology* **79**, 903–7.

Sheppard J.W., Wright P.F. & Desavigny D.H. (1986) Methods for the evaluation of EIA tests for use in the detection of seedborne diseases. *Seed Science and Technology* **14**, 49–59.

Shibata D.K., Arnheim N. & Martin J.W. (1988) Detection of human papilloma virus in paraffin-embedded tissue using the polymerase chain reaction. *Journal of Experimental Medicine* **167**, 225–30.

Stead D.E. & Pemberton A.W. (1987) Recent problems with *Pseudomonas syringae* pv. *pisi* in the UK. *Bulletin OEPP/EPPO* **17**, 291–4.

Taylor J.D. (1970) The quantitative estimation of the infection of bean seed with *Pseudomonas phaseolicola* (Burkh.) Dowson. *Annals of Applied Biology* **66**, 29–36.

Taylor J.D. (1984) In *Report on the 1st International Workshop on Seed Bacteriology*, 1982, Angers, pp. 9–14. International Seed Testing Association, Zurich.

Taylor J.D., Dudley C.L. & Presly L. (1979) Studies of halo-blight seed infection and disease transmission in dwarf beans. *Annals of Applied Biology* **93**, 267–77.

Taylor J.D., Bevan J.R., Crute I.R. & Reader S.L. (1989) Genetic relationship between races of *Pseudomonas syringae* pv. *pisi* and cultivars of *Pisum sativum*. *Plant Pathology* **38**, 364–75.

Tenover F.C. (1988) Diagnostic deoxyribonucleic acid probes for infectious diseases. *Clinical Microbiology Reviews* **1**, 82–101.

van Vuurde J.W.L. & Maat D.Z. (1983) Routine application of ELISA for the detection of lettuce mosaic virus in lettuce seeds. *Seed Science and Technology* **11**, 505–13.

van Vuurde J.W.L. & Maat D.Z. (1985) Enzyme-linked immunosorbent assay (ELISA) and disperse dye immunoassay (DIA): comparison of simultaneous and separate incubation of sample and conjugate for the routine detection of lettuce mosaic virus and pea early browning virus. *Netherlands Journal of Plant Pathology* **91**, 3–13.

van Vuurde J.W.L., van den Bovenkamp G.W. & Birnbaum Y. (1983) Immunofluorescence microscopy and enzyme-linked immunosorbent assay as potential routine tests for the detection of *Pseudomonas syringae* p.v. *phaseolicola* and *Xanthomonas campestris* pv. *phaseoli* in bean seed. *Seed Science and Technology* **11**, 547–59.

Varveri C., Candresse T., Cugusi M., Ravelonandro M. & Dunez J. (1988) Use of a ^{32}P-labelled transcribed RNA probe for dot hybridisation detection of plum pox virus. *Phytopathology* **78**, 1280–3.

Ward E.S., Güssow D., Griffiths A.D., Jones P.T. & Winter G. (1989) Binding activities of a repertoire of single immunoglobulin variable domains selected from *Escherichia coli*. *Nature* **341**, 544–6.

12 From the research bench to the market place: development of commercial diagnostic kits

S.A. MILLER, J.H. RITTENBURG,
F.P. PETERSEN & G.D. GROTHAUS

Introduction

The ability of crop managers, consultants and plant health professionals to detect and quantify plant pathogens in crops, soil and water is improving rapidly, primarily as a result of the development and application of new techniques in immunology, immunochemistry and molecular biology. The technology is now available to diagnose specific diseases rapidly and accurately, and to monitor the spread of a pathogen or disease in a crop, permitting more rational disease management. Monoclonal antibody technology was introduced in 1975 by Kohler & Milstein and applied to plant viruses in 1982 (Halk et al., 1982; Dietzgen & Sander, 1982); it has now been applied to many groups of plant pathogens and pests (Jones et al., 1988; Miller & Martin, 1988; Schots et al., 1989). This has resulted in the production of high quality, consistent, specific immunoreagents for detection of a wide range of plant pathogens. Also, improvements in purification methods now allow production of polyclonal antisera with enhanced specificity and sensitivity (Weir et al., 1986). Antibodies coupled with improved assay technology can now be used to detect and quantify pathogens 'on-site'. Commercial products are available from a handful of companies for the detection of viruses, mycoplasmas, bacteria and fungi. These products range from laboratory assays requiring several hours or days to complete to tests that can be used in the field and give results in 10 min. All of the commercial assays to date utilize immunoassays. While DNA probes are being developed for many micro-organisms in research laboratories throughout the world, assays utilizing these reagents do not at this time approach the sensitivity, rapidity, ease-of-use and cost-effectiveness of immunoassays. Therefore, this chapter is necessarily focused on immunoassay technology and its application to the detection of fungal pathogens of plants.

Steps in immunoassay development

Plant pathogens present unique challenges for immunoassay development, from concept through to field use and analysis.

Some of these include the following.

1 Presence of numerous strains, races and pathovars.

2 Lack of information on biochemical characteristics of many complex pathogens, such as fungi and nematodes.

3 Wide range of crops, crop varieties and environmental conditions under which they are grown.

4 Potential interference from plant sap and soil components.

5 Low concentrations of pathogens in some crops and soil.

6 Lack of aetiological and/or epidemiological data that can be readily related to immunoassay results.

Antibody production and selection

Most plant pathogenic fungi have numerous stages in their life cycles, and complete information on the critical life stage(s) to target in an immunoassay may not be available. It may be necessary to limit detection to a particular life stage, or to detect several very different life stages. For example, an assay may be required to detect mycelial antigens in plant tissue and also resting structures of the same pathogen in soil. This can be a problem with fungal pathogens because so little is known about the biochemical composition of antigens of these fungi. There is no way of knowing, *a priori*, the proportion and type of antigens shared by different life stages of the pathogen.

Once a life stage(s) is chosen, it is produced in culture or, in the case of obligate parasites, in the plant and purified for antibody production. Antigens are extracted from the pathogen and crude or purified extracts used to immunize animals. Antigens may be extracted by cell disruption (e.g. freeze-drying/grinding), cell washing or other methods (Boonekamp, 1988; Dewey & Brasier, 1988).

The method of preparing the immunogen must take into account the format of the immunoassay that will be used. Most commercial assays for plant pathogens are based on double-antibody enzyme-linked immunosorbent assay (dabELISA), which is ideally suited for the detection of antigens in complex mixtures such as plant sap and soil extracts. Intact bacteria, spores and mycelial cells are too large to bind to the primary capture antibody in a dabELISA; however, soluble antigens secreted by the cells or released by washing or cell disruption will be detected (Dewey, this volume). The immunogen should be prepared in a way that maximizes the concentration of soluble antigens if these antigens are to be detected in the final assay.

Both monoclonal and polyclonal antibodies have been used successfully in immunoassays for plant pathogens (Amouzou-Alladaye *et al.*, 1988; Boonekamp, 1988; Dewey & Brasier, 1988; Miller & Martin, 1988; Petersen *et al.*, 1989). Polyclonal antisera can be produced in large lots, absorbed against cross-reactive antigens if necessary, and purified to provide highly sensitive, consistent reagents. However, polyclonal antisera may not have the required specificity to detect some fungi and other complex micro-organisms; in such cases monoclonal antibodies become the reagent of choice. Because a monoclonal antibody reagent contains only a single type of antibody, it is possible to select one or more that have excellent specificity for the target pathogen. Developing an immunoassay in which monoclonal antibodies are used is not always straightforward, and mono-clonal antibodies often cannot be handled in the same way as polyclonal antiserum. Hybridomas isolated in a typical fusion experiment secrete monoclonal antibodies exhibiting a range of characteristics, including epitope specificity and affinity, immunoglobulin class and subclass, pH sensitivity and ability to be conjugated to an enzyme (Goding, 1983; Vaidya *et al.*, 1985; Boonekamp, 1988).

The final format of the assay to be used in the field or laboratory should be considered and incorporated into the antibody screening/evaluation process as early as possible. A mono-clonal antibody selected on the basis of reactivity with a target pathogen in an indirect ELISA in which the antigen is bound to the microtitre plate may not function well as the capture or conjugated antibody in dabELISA.

The specificity of monoclonal antibodies for particular epitopes requires that they be screened against a large number of related and unrelated micro-organisms. This requires building and maintaining a large collection of isolates, developing methods that optimize growth in culture, growing the isolates and pre-paring standardized extracts for screening. The number of isolates that should be screened varies, depending on morpho-logical variation in the genus, species or strain to be detected, and the degree of relatedness to other micro-organisms. Mono-clonal antibodies that cross-react with non-target microbes are eliminated from consideration at an early stage. Highly specific monoclonal antibodies, which react only with a portion of the target isolates, may be combined with other monoclonal anti-bodies with complementary specificity (Mitchell & Sutherland, 1986; De Boer & McNaughton, 1987).

Sample preparation   One of the challenges of developing immunoassays for field use is to find simple, efficient, convenient methods of extracting pathogens from plant tissue or other matrices. We have devel-oped a simple tissue-grinding device for extracting fungal anti-gens from turfgrass and other crops. The Extrak™ Sample Preparation Pad consists of three strips of abrasive material attached to a small plastic card. A plant sample is placed on the pad and ground thoroughly with a square pad of the same abrasive material. Once the pads are saturated with plant sap, the three small strips are removed from the plastic card and placed in extraction buffer in a bottle. A filter placed in the bottle neck removes components of plant sap that interfere with the test. This method allows a defined amount of plant extract to be added to the buffer for consistent extraction without the need to weigh the sample. Samples can also be ground effectively in a defined amount of buffer in a microcentrifuge tube using a matched plastic pestle, in the field or in laboratories with minimal equipment. Another simple method of extracting virus from plant samples, suitable for the field, is the squash-blot method in which leaves are squashed between two strips of nitrocellulose attached to plastic backing. The sap is dried and the strips are evaluated in the laboratory by ELISA (Mitchell *et al.*, 1988).

Methods of extraction that we have found to be suitable for use in research laboratories or in diagnostic clinics and that are geared towards small volumes of sample include: (i) disruption using an inexpensive cell disrupter employing a plastic capsule and metal or plastic ball pestle (Dental Darby Inc.); and (ii) grinding samples frozen in liquid nitrogen in a standard mortar and pestle.

Immunoassay characteristics

Agri-Diagnostics Associates have produced immunoassay kits for the detection of fungal and bacterial plant pathogens. The assays, in 'research' and 'on-site' formats, are based on dabELISA and employ monoclonal and/or polyclonal antibodies. Dipstick assays for the diagnosis and monitoring of rhizoctonia brown and yellow patch (*R. solani* and *R. cerealis*), pythium blight (*Pythium aphanidermatum* and other *Pythium* spp.), and dollar spot (*Sclerotinia homoeocarpa*) in golf course turfgrass were introduced commercially in 1987 (Miller *et al.*, 1989 a), and 10-min. 'on-site' assays for these diseases were introduced in 1989 (Rittenburg *et al.*, 1988; Miller *et al.*, 1988 b). The 'on-site' immunoassay format is specially designed for speed, simplicity, specificity and sensitivity. The assay utilizes a 'flow-through' design in which the capture antibody is immobilized on the surface of an absorbant material; the sample extract, enzyme-conjugated second antibody, rinse solution, enzyme substrate and finishing solution are added sequentially from dropper bottles. The enzyme conjugated to the second 'tag' antibody is horseradish peroxidase, and the substrate is 4-chloro-naphthol. As the substrate flows through the reaction area a blue precipitate will form if any bound peroxidase is encountered. If no pathogen was present in the sample, no colour formation will occur (Fig. 12.1). The amount of colour that develops in a positive reaction is proportional to the amount of pathogen present in the sample. The assay includes internal negative and positive controls, and results can be read with a simple hand-held densitometer. The kit is entirely self-contained and requires no mixing or measuring of reagents. A similar assay has also been developed for the detection of *Phytophthora* in plant tissue (Miller *et al.*, 1989 b).

Research assays are also available for detection of fungi in the genera *Phytophthora*, *Pythium*, *Rhizoctonia* and *Sclerotinia*, and for the geranium bacterial wilt pathogen, *Xanthomonas campestris* pv. *pelargonii*. These laboratory assays are configured as dabELISAs in 96-well microtitre plates (Fig. 12.2).

The wells are pre-coated with capture antibody and assays are done by adding sample extract, enzyme-conjugated antibody and substrate in sequence, with a washing step after incubation of sample extract and conjugate. The sample, conjugate and substrate are incubated for 10 min each with shaking. A stop solution is added after the substrate incubation to inhibit further colour development. The assay can then be read on any standard microtitre plate reader. The entire test, from sample loading through to colorimetric analysis, can be completed in 45−60 min. The kits are well suited for analysis of large numbers of samples and are particularly useful for research studies. They are used extensively in our own work to evaluate antibodies for sensitivity and specificity and to develop information relative to pathogen levels in crops and soil.

Fig. 12.1 Rapid immunoassay devices for plant pathogens. On each device the lower right well is the negative control well while the lower left well is the sample well. The small well is an internal positive control. The device on the left shows a strong positive reaction while the device on the right shows a typical negative result.

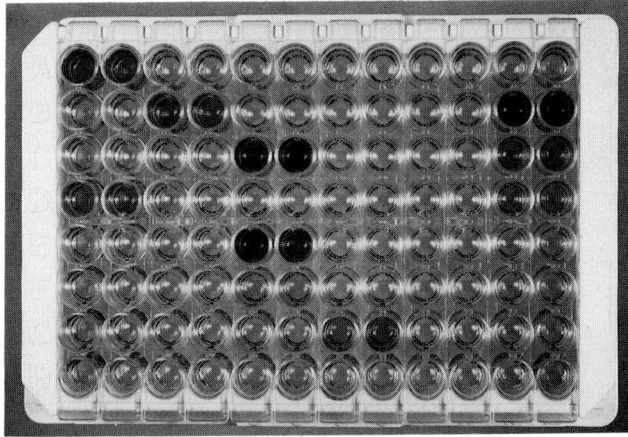

Fig. 12.2 Double-antibody multiwell ELISA for detection of *Phytophthora* spp.

Dose—response curves for the rhizoctonia brown patch and pythium blight rapid disease detection kits (Figs 12.3 & 12.4) demonstrate that the kits have sensitivity thresholds of approximately 0.5 and $0.05\,\mu g$ fungal protein ml^{-1}, respectively. A logarithmic response curve with a dynamic range of approximately two orders of magnitude before reaching a plateau at substrate density saturation was observed in both assay systems. No non-specific matrix effect from turfgrass foliage was observed.

The rhizoctonia assay is highly specific for *R. solani* and *R. cerealis* (Table 12.1). *Rhizoctonia zeae* and *R. oryzae* reacted weakly in the assay compared to the other *Rhizoctonia* spp. Other genera of fungi including *Typhula*, *Coprinus*, *Sclerotinia*, *Fusarium*, *Curvularia*, *Pythium* and *Rhizopus* also reacted only weakly in the assay. The specificity of the antibodies used in this assay was checked against a larger panel of isolates in a dabELISA (multiwell format), including 24 isolates of *Rhizoctonia solani*, 19 isolates of *R. cerealis*, and other fungi associated with turfgrass, with results similar to those observed using the rapid format.

In cross-reactivity tests against a wide range of *Pythium* species, the pythium rapid assay demonstrated good reactivity with all of the species of *Pythium* known to cause turfgrass disease (Table 12.2). Non-oomycetous fungi are not detected in this assay but reactivity with some *Phytophthora* spp. has been observed. However, *Phytophthora* spp. have never been reported to occur on turfgrass and reactivity of the antibodies with these pathogens does not affect field performance of the kit.

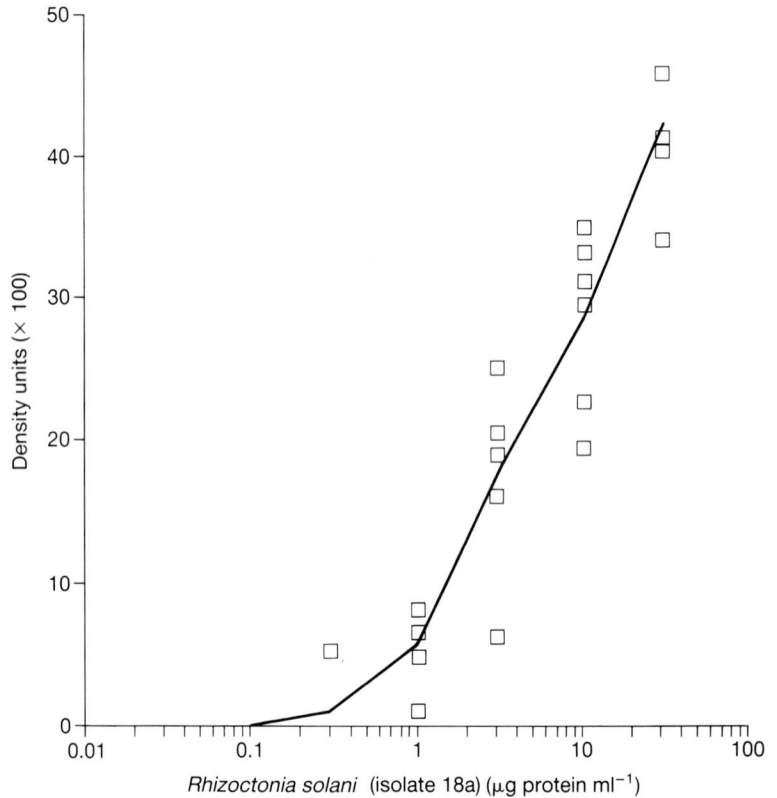

Fig. 12.3 Dose—response curve for the rhizoctonia rapid immunoassay, prepared using extracts of a pure culture of *Rhizoctonia solani*. Results were quantified using a handheld reflectometer (AgriMeter II, Agri-Diagnostics Associates).

Table 12.1 Reactivity of the rhizoctonia rapid assay with extracts of pure cultures of fungal pathogens. All extracts were standardized at a protein concentration of $10\,\mu g\ ml^{-1}$. Density units were determined with a Gretag Model D-152 reflectometer.

Species	Number of isolates tested	Mean density units ($\times100$)
Rhizoctonia solani	4	46.5
Rhizoctonia cerealis	2	32.5
Rhizoctonia oryzae	1	14.0
Rhizoctonia zeae	2	3.5
Typhula spp.	2	8.0
Coprinus psychromorbidus	1	6.0
Sclerotinia homoeocarpa	1	5.0
Curvularia spp.	1	2.0
Fusarium nivale	1	2.5
Pythium aphanidermatum	1	0.5
Rhizopus stolonifer	1	1.0

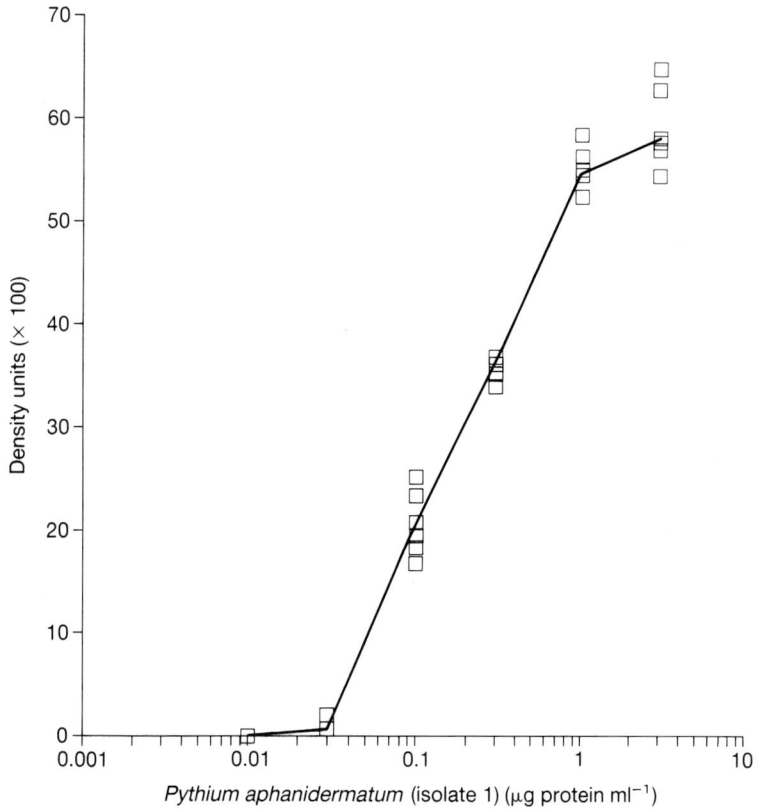

Fig. 12.4 Dose−response curve for the pythium rapid immunoassay, prepared using extracts of a pure culture of *Pythium aphanidermatum*. Results were quantified using a handheld reflectometer (AgriMeter II, Agri-Diagnostics Associates).

Table 12.2 Reactivity of the pythium rapid assay with extracts of pure cultures of fungal pathogens. All extracts were standardized at a protein concentration of $10\,\mu g\ ml^{-1}$. Density units were determined with a Gretag Model D-152 reflectometer.

Pythium sp.	Causes of disease in turfgrass	Number of isolates tested	Mean density units ($\times 100$)
aphanidermatum	+	5	54.8
ultimum	+	2	47.0
myriotylum	+	2	42.5
graminicola	+	2	38.5
irregulare	+	2	38.0
arrhenomanes	+	2	37.0
heterothallicum	−	1	44.0
splendens	−	2	32.0
salpingophorum	−	1	51.0
sylvaticum	−	1	0.0
dissotochum	−	1	1.0
coloratum	−	1	2.0

Field results

One critical application of rapid diagnostic kits is in monitoring crops for changes in pathogen populations over a period of time. For many pathogens, particularly those that do not produce airborne spores, this has not been possible using conventional techniques. However, the biomass of a pathogen can be measured in samples using quantitative immunoassays, thus providing information relating to population shifts with time as environmental conditions change and/or fungicides are applied. We carried out a monitoring study on a New Jersey golf course to evaluate the ability of the rapid immunoassay to quantify populations of *Rhizoctonia* in 6 mm bentgrass golf greens (Miller *et al.*, 1988 b). Three greens were chosen on the basis of disease history: symptoms of rhizoctonia brown patch were observed in previous years on greens 12 and 14, while green 16 had not shown symptoms of the disease. The study was started in May 1988 and continued until August; samples were collected and tested at approximately weekly intervals after 7 June. Six samples from each of the three greens were mixed thoroughly, and two subsamples were removed and tested in the rhizoctonia rapid immunoassays. Eight blades of turfgrass were also selected from each sample and placed on acid water agar for isolation of *Rhizoctonia solani*. The results of the monitoring programme are presented in Fig. 12.5 and Table 12.3.

On green 12, an initial low positive immunoassay reading was observed early in May (May 13, day 0) in the absence of symptoms of brown patch and an unidentified species of *Rhizoctonia* was isolated from the grass blades (Table 12.3). Rapid immunoassay results were negative until day 53 (6 July) but began to increase prior to the development of symptoms, which were observed on day 69 (22 July). The highest densitometer values were obtained at the peak of symptom development and decreased as the disease severity declined. The disease was active on days 69 and 73 and remnants of patches were visible on day 80. Symptoms were no longer present at the last sampling date and immunoassay results for samples taken at that time were negative. *Rhizoctonia solani* was only isolated on day 69, when active disease development was first observed.

Results of the monitoring program for green 14 were similar to those of green 12 (Fig. 12.5). Brown patch was observed a few days later on green 14 than on green 12 and this was reflected in the immunoassay results. No symptoms were observed on green 16; no *Rhizoctonia* spp. were isolated and immunoassay results were consistently negative.

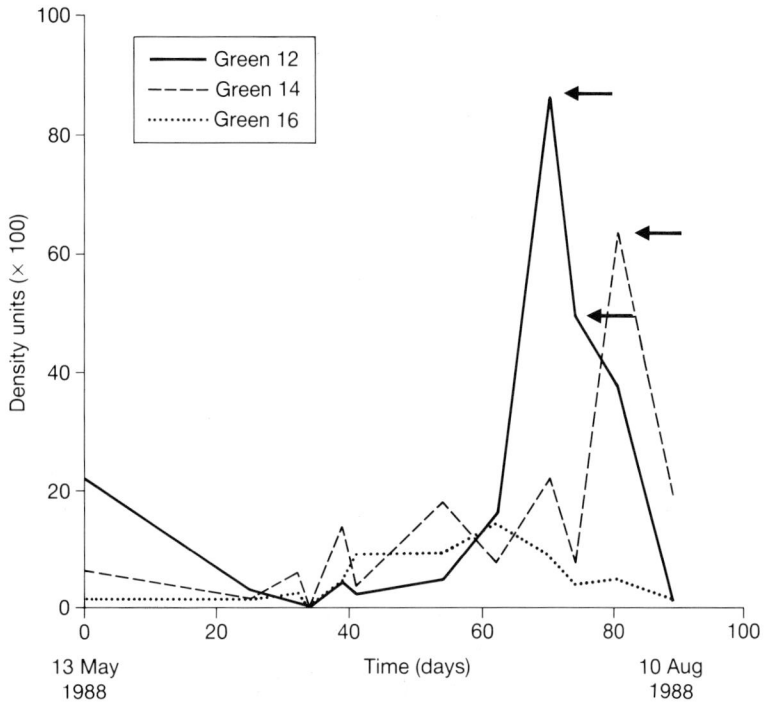

Fig. 12.5 Monitoring of rhizoctonia brown patch on golf greens in New Jersey, USA in 1988 using rhizoctonia rapid immunoassays. Results were quantified using a Gretag Model D-152 reflectometer. Arrows indicate appearance of brown patch symptoms in the turfgrass.

Table 12.3 Isolation of *Rhizoctonia* sp. from golf course greens in southern New Jersey, USA

| Sampling day* | Isolates of *Rhizoctonia* sp. recovered from | | |
	Green 12	Green 14	Green 16
0	1 Rsp	1 Rs	0
25	1 Rz	1 Rsp	0
32	0	0	0
34	0	0	0
39	0	0	0
45	0	0	0
53	0	0	0
61	0	1 Rz	0
69	1 Rs/1Rsp	0	0
73	0	3 Rs	0
80	0	0	0
88	0	0	0

* Day 0 = May 13, 1988; day 88 = August 10, 1988.
Rsp = *Rhizoctonia* sp.; Rs = *R. solani*; Rz = *R. zeae*.

Pythium blight   Pythium blight is a potentially devastating disease that occurs in cool season grasses in the United States primarily in July and August, when overnight temperature and relative humidity are high. Severe damage can occur in less than 1 day if environmental conditions are favourable for the disease and fungicides are not applied (Smiley, 1983). Active pythium blight is characterized by the presence of fluffy white mycelium over the surface of the turf, as a result of its spread from leaf to leaf. Several turfgrass diseases may also have white mycelium and thus, diagnosis may not always be straightforward. In addition, it is preferable to control the disease before the mycelia are visible and the turfgrass foliage is damaged.

Studies carried out in 1988 in turfgrass field plots demonstrated that *Pythium* spp. could be detected at an early stage of disease development using the pythium rapid immunoassays (Miller *et al.*, 1988 a). *Pythium* spp. were detected in chlorotic perennial ryegrass foliage as well as in dead patches, but not in foliage that appeared to be healthy (Fig. 12.6).

Disease development was tracked over a 2-week period in the perennial ryegrass plot (Fig. 12.7). Immunoassay meter readings were high in the symptomatic area when symptoms were present but were much smaller when the disease subsided in the last

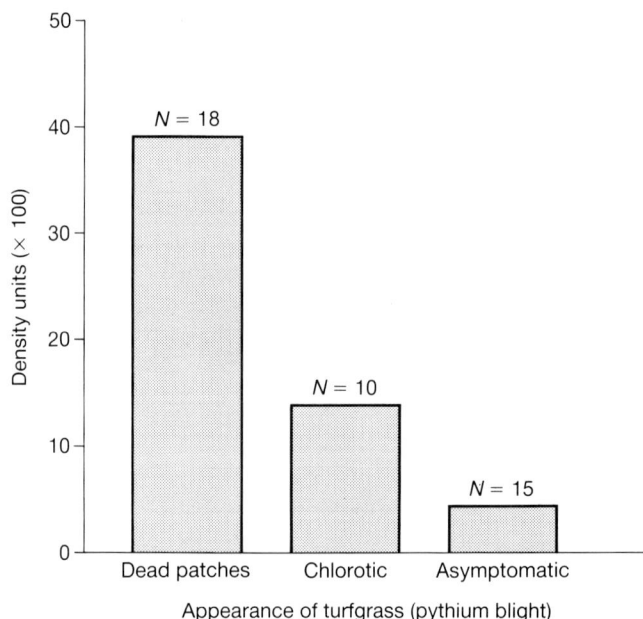

Fig. 12.6 Detection of pythium blight in perennial ryegrass samples by means of the pythium rapid immunoassay. Results were quantified using a Gretag Model D-152 reflectometer. Samples were taken (N = number of samples) from the margin of necrotic, chlorotic or asymptomatic areas.

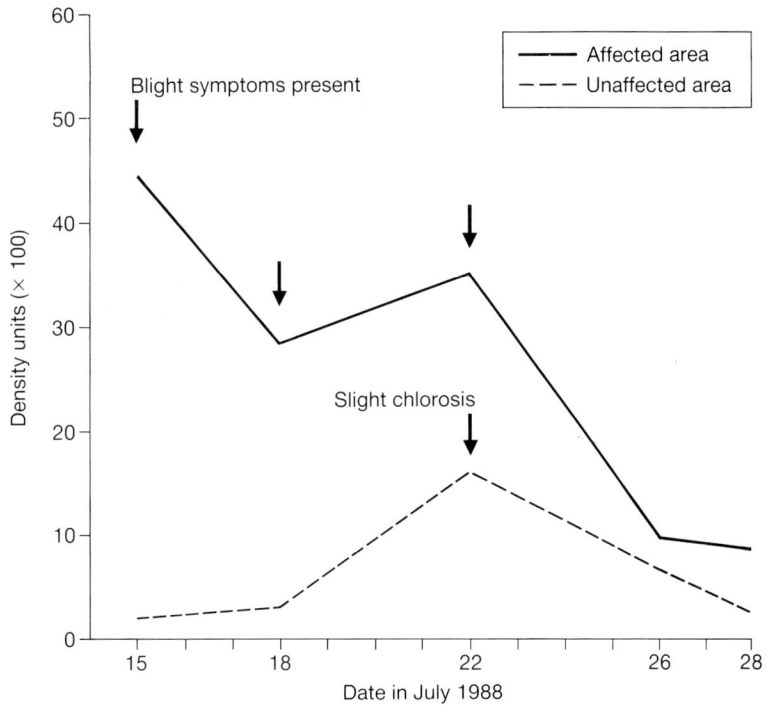

Fig. 12.7 Monitoring of pythium blight in perennial ryegrass using pythium rapid immunoassays. Results were quantified using a Gretag Model D-152 reflectometer. No symptoms were present in the asymptomatic area except on July 22 when slight chlorosis was observed.

two sampling periods of the study. Density values remained small in the grass samples from the asymptomatic area, except for the July 22 sample, when slight chlorosis was observed and *Pythium* spp. were detected. Tests on golf course greens and fairways have shown similar results (Miller *et al.*, 1989 a). The rapid immunoassays thus provide golf course superintendents with a means of accurately diagnosing pythium blight and detecting it at an early stage of development so that proper control practices can be implemented.

Conclusions

The results presented above indicate that immunoassays can be used to monitor population changes of fungal pathogens in a crop over a period of time. This information is useful not only in making an accurate diagnosis but also in early detection of pathogen population increases. Knowledge of pathogen activity within a crop can be used to optimize applications of fungicides for maximum effectiveness or implementation of other control measures. Such information will become increasingly important as crop managers seek to produce crops competitively and to minimize environmental impacts.

The technical challenges inherent in developing user-friendly, specific, sensitive assays to detect pathogens in infected plant tissue have been met and products are being made available in agricultural markets. The next generation of products will include assays for detection of soil-borne plant pathogens. For such products to be useful, technical issues such as elimination of effects of soil, collection and concentration of propagules, and sample preparation must be addressed successfully. However, these technical problems can be solved, as has been demonstrated by the development of an immunoassay for detection of *Phytophthora* in field soils (Miller *et al.*, 1989 b).

Immunoassay technology has come a long way for agricultural applications in a relatively short time. A decade ago, immunoassays were generally considered useful but complicated and time-consuming processes that could only be carried out by highly trained scientists in modern research laboratories. Today, immunoassays are being carried out in the field by golf course superintendents and others without technical training, who are using the information provided by the tests to make crop management decisions. In the future, agricultural production will benefit from the availability of immunoassays for the detection of plant pathogens affecting a wide range of crops.

References

Amouzou-Alladaye E., Dunez J. & Clerjeau M. (1988) Immunoenzymatic detection of *Phytophthora fragariae* in infected strawberry plants. *Phytopathology* **78**, 1022–6.

Boonekamp P.M. (ed.) (1988) *Monoclonal Antibodies and Immunological Techniques to Detect Plant Pathogens*. Proceedings of the 1st COST-88 Workshop, Wageningen, 24–27 November, 1987. Pudoc, Wageningen.

De Boer S.H. & McNaughton M.E. (1987) Monoclonal antibodies to the lipopolysaccharide of *Erwinia carotovora* subsp. *atroseptica* serogroup I. *Phytopathology* **77**, 828–32.

Dewey F.M. & Brasier C.M. (1988) Development of ELISA for *Ophiostoma ulmi* using antigen-coated wells. *Plant Pathology* **37**, 28–35.

Dietzgen R.G. & Sander E. (1982) Monoclonal antibodies against a plant virus. *Archives of Virology* **74**, 197–204.

Goding J.W. (1983) *Monoclonal Antibodies: Principles and Practice*. Academic Press, London.

Halk E.L., Hsu H.T. & Aebig J. (1982) Properties of virus-specific monoclonal antibodies to *Prunus* necrotic ringspot, apple mosaic, tobacco streak and alfalfa mosaic viruses. *Phytopathology* **72**, 953.

Jones P., Ambler D.J. & Robinson M.P. (1988) The application of monoclonal antibodies to the diagnosis of plant pathogens and pests. *Proceedings of the Brighton Crop Protection Conference – Pests and Diseases 1988*, Vol. 2, pp. 767–76. British Crop Protection Council, Thornton Heath, Surrey.

Kohler G. & Milstein C. (1975) Continuous culture of fused cells secreting antibody of predefined specificity. *Nature* **256**, 495–7.

Miller S.A. & Martin R.R. (1988) Molecular diagnosis of plant disease. *Annual Review of Phytopathology* **26**, 409–32.

Miller S.A., Plumley K.A., Rittenburg J.H., Petersen F.P. & Grothaus G.D. (1988 a) Rapid detection and monitoring of Pythium blight of turfgrass by means of a field usable enzyme-linked immunoasorbent assay (ELISA). *Phytopathology* **78**, 1591 (Abs.).

Miller S.A., Rittenburg J.H., Petersen F.P. & Grothaus G.D. (1988 b) Application of rapid, field-usable immunoassays for the diagnosis and monitoring of fungal pathogens in plants. *Proceedings of the Brighton Crop Protection Conference — Pests and Diseases 1988*, Vol. 2, pp. 795–803. British Crop Protection Council, Thornton Heath, Surrey.

Miller S.A., Grothaus G.D., Petersen F.P. & Rittenburg J.H. (1989 a) Detection and monitoring of turfgrass pathogens by immunoassay. In Leslie A.R. & Metcalf R.L. (eds) *Pesticide Problems and IPM Solutions for Urban Turfgrass and Ornamentals*, pp. 109–20. U.S. Environmental Protection Agency, Office of Pesticide Programs, Washington D.C.

Miller S.M., Petersen F.P., Miller S.A., Rittenburg J.H., Wood S.C. & Grothaus G.D. (1989 b) Development of a direct immunoassay to detect *Phytophthora megasperma* f.sp. *glycinea* in field soil. *Phytopathology* **79**, 1139 (Abs.).

Mitchell L.A. & Sutherland J.R. (1986) Detection of seedborne *Sirococcus strobilinus* with monoclonal antibodies in an enzyme-linked immunosorbent assay. *Canadian Journal of Forestry Research* **16**, 945–8.

Mitchell D.H., Rose D.G. & Howell P.J. (1988) European Patent Application No. 88300937.5, Squash Blot Device, European Patent Office, Munich.

Petersen F.P., Maybroda A.M. & Miller S.A. (1989) Monoclonal antibodies to *Sclerotinia homoeocarpa*. *United States Patent* No. 4 803 155. United States Patent Office, Washington D.C.

Rittenburg J.H., Petersen F.P., Miller S.A. & Grothaus G.D. (1988) Development of a rapid, field-usable immunoassay format, for detection and quantitation of *Pythium*, *Rhizoctonia* and *Sclerotinia* spp. in plant tissue. *Phytopathology* **78**, 1519 (Abs.).

Schots A., Hermsen T., Schouten S., Gommers F. & Egberts E. (1989) Serological differentiation of the potato-cyst nematodes *Globodera pallida* and *G. rostochiensis*: II. Preparation and characterisation of species specific monoclonal antibodies. *Hybridoma* **8**, 401–11.

Smiley R.W. (1983) *Compendium of Turfgrass Diseases*. American Phytopathological Society, St. Paul, Minnesota.

Vaidya H.C., Dietzler D.N. & Ladenson J.H. (1985) Inadequacy of traditional ELISA for screening hybridoma supernatants for murine monoclonal antibodies. *Hybridoma* **4**, 271–6.

Weir D.M., Herzenberg L.A., Blackwell C. & Herzenberg L.A. (1986) *Handbook of Experimental Immunology*, 4th edn, Vol. 1. Blackwell Scientific Publications, Oxford.

13 Commercialization of serological tests for plant viruses

P. GUGERLI

Introduction

Commercial interest in the serological identification and detection of plant viruses, and in diagnosis of the diseases they cause, started about 10 years ago. The principal stimulus was the introduction of the very sensitive and versatile enzyme-linked immunosorbent assay (ELISA) by Voller *et al.* (1976) and Clark & Adams (1977), which made possible many new applications for serological techniques, especially the large-scale, routine testing of plants. It opened up a potential market of a critical minimum mass (Fig. 13.1), not just by increasing the number of applications but also through the higher, added value of the new reagents; purified and concentrated immunoglobulins and the corresponding enzyme conjugates could be used at much greater dilution in ELISA, and were therefore of higher monetary value than the crude antisera used in simple immunoassays in the past. The introduction of monoclonal antibodies (Dietzgen & Sander, 1982; Gugerli 1982; Gugerli & Fries, 1983) was a further impetus to commercialization, but other factors also hastened the process: the possibility of transferring know-how from the medical to the agricultural diagnostic field, and the rapid development of plant related biotechnologies. Cutbacks in funding and the trend towards transferring some of the cost of public research to industry also increased awareness of commercial possibilities within government-funded institutes. One of the first companies to sell antisera and thereby initiate the new commercial era was the Swiss-based company Bioreba A.G. (formerly Inotech Diagnostik A.G.) working in collaboration with the Federal Agricultural Research Station of Changins, Nyon.

From the manufacturer to the user

Commercialization requires the development, manufacture, distribution and application of reliable immunological reagents. The basic research preceding development is not discussed here, since it is, in the main, still carried out independently by public research institutes.

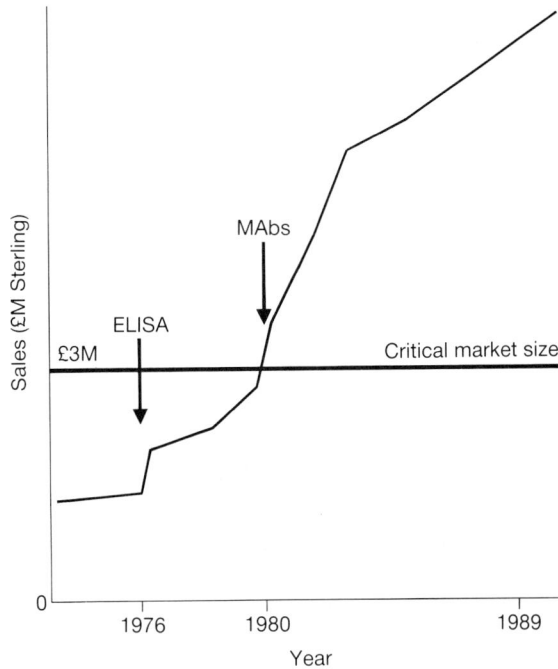

Fig. 13.1 Estimated value of the diagnostic market after introduction of ELISA and monoclonal antibodies (MAbs).

Development includes pilot production, quality optimization, standardization, design of the kit, elaboration of the application protocol and large-scale verification of it under conditions in which it will be used commercially. Manufacture covers the scaling-up of production, quality control and storage of the final diagnostic product. Distribution is as complex as production and involves marketing, advertising, product presentation, international distribution, trade agreements and formalities, service and training, all of which are costly. At the end of the line are the consumers who have to pay for the product (and indirectly all the stages leading to its production). Consequently, they are likely to be more demanding than the traditional users of free reagents from the scientific community. They will expect to get a standardized product, with instructions and advice for its application, backed up where necessary by customer service to help them with any particular problems which they may encounter in using the test. For the manufacturer, the customer's response is the driving force within the product development process.

Manufacturers and their products

Six companies share the market for products for diagnosing

plant viruses (Table 13.1). Despite the very limited market at present, several new companies have declared an interest in this field over the last few years. Together they offer reagents for the immuno-enzymatic detection of almost 100 different viruses. Reagents for viruses of vegetatively propagated plants such as potato, fruit trees, grapevine and flowers predominate. The majority are offered in the form of simple kits containing purified immunoglobulins and enzyme−immunoglobulin conjugates for either 50, 100, 250, 500, 1000 or 5000 tests. Some kits also contain special buffers and a few include all buffers, solid-phase supports and positive and negative reference samples. So far most kits have been for laboratory use and simple, rapid and sensitive tests for field applications are not yet available. Bioreba have sold monoclonal antibodies for the detection of potato viruses since 1982. In addition to the antisera and complete kits, they also offer buffers and solid-phase supports separately, as well as equipment, some of it specially designed for the routine extraction of plant tissue and handling liquids (Fig. 13.2).

The cost per test of kits ranges from £0.03 to £1.80, which is considerably cheaper than similar tests used in medicine. Prices vary considerably between similar products from different manufacturers, reflecting the rapidly developing nature of the market.

Table 13.1 Manufacturers of reagents for enzyme immunoassay kits

Company	Country of origin	When started	Number of products	Number of tests/kit	Price per test (£)
Bioreba AG*	Switzerland	1980	26 (+60)	500	0.42−0.53
				1000	0.14
				5000	0.04−0.05
Agdia Inc.*	USA	1981	60 (+26)	288	0.29−0.41
				480	0.27−0.39
				1000	0.11−0.21
Boehringer Mannheim GmbH	Germany	1982	16	50	1.80
				250	0.52
				500	0.30−0.31
				1000	0.16
Ingenasa	Spain	1983	4	5000	0.03−0.12
Sanofi Sante Anim. SA	France	1986	24	100	0.46−0.55
				500	0.09−0.11
				1000 kit	0.20−0.27
Loewe	Germany	1988	46	500	0.15−0.25

* Bioreba AG and Agdia Inc. have a reciprocal agreement. These companies also sell monoclonal antibody-based test kits. Bioreba also sells buffers, disposables and instrumentation.

Fig. 13.2 Tecan plant sap extractor developed by P. Gugerli.

The market

The market for diagnostic products for plant viruses is primarily defined by the impact of the virus diseases on the agricultural or horticultural crops, as judged by the commercial profit obtained with virus-free *vis à vis* virus-infected plants. Disease propagation and virus-transmission modes, as well as crop establishment costs and crop duration, are important parameters in determining market viability for a product. The highest price per test is paid for phytosanitary applications in basic multiplication schemes producing planting material of perennial crops such as grapevine. However, the health status of such long-lived crops is checked before field establishment, and testing is not required annually, which constrains the size of the market. In contrast, phytosanitary control of annual crops like potato needs much more frequent inspections of the crop, and the numbers of tests are considerably higher.

The market is also directly influenced by the value of the agricultural products which, in turn, is governed by commercial, social and political constraints. There are large differences in priorities between different agricultural communities and in their agricultural and phytosanitary standards and practices.

The structure of the market will therefore vary considerably between different geographical or political regions. The most important customers are generally large co-operative organizations of farmers, often supported by their respective governments and official quarantine and inspection services. Research virologists are also a valuable class of customers, as are plant breeders and dealers in plant propagation material and agricultural products. Lastly, industrial-type farms and nurseries have an increasing interest in simple reliable kits, especially for routine service testing.

The phytodiagnostic market differs considerably from the medical diagnostic market. There are no health insurance systems to cover the costs of plant disease diagnosis and the most frequent phytodiagnostic applications concern populations, which contrasts with medicine where the tests are most frequently done on individuals. A diagnostic test for a plant virus disease adds a commercially assessable added value to the tested crop. In all, the phytodiagnostic market is slightly more market-driven than its medical counterpart, a view confirmed by the lower prices of phytodiagnostic reagents.

Evolution of the market

The market for diagnosing plant diseases is dominated by virological applications. It is, however, still very small, having an estimated turnover of less than £10 million in 1989 (Fig. 13.3). It is therefore surprising that six companies have invested in it. The potential market is much larger, but political and social factors such as state intervention through publicly funded agricultural research, the provision of free advisory services, and a lack of money and slow technology transfer in Third World countries are hampering its development. If, for example, the same standards for potato seed certification which operate in some of the European potato-producing countries were applied worldwide, the potato virus reagent market would, on its own, be more important than the present market for all phytodiagnostic products (Table 13.2). At the moment, any companies investing in this field must be doing so either for the prestige, the hope of lucrative monopolies in the future, or because of poor market analysis.

Advantages

The commercial application of serological techniques for plant viruses should result in ready access to reliable diagnostic reagents. Competition should improve quality, and large-scale production yields standardized products which leads to more comparable test results between large numbers of laboratories.

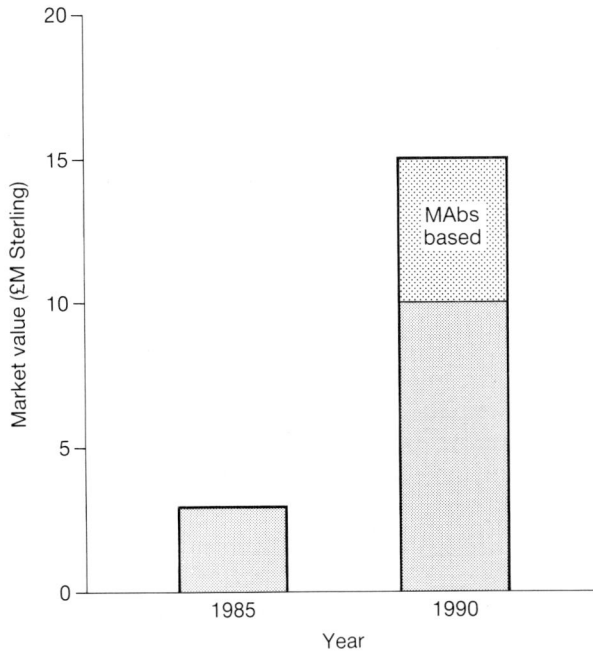

Fig. 13.3 Development of the phytodiagnostic market (MAbs = monoclonal antibodies).

Table 13.2 World potato production and costs of ELISA reagents for a theoretical worldwide potato seed certification programme based on Swiss practices and costs of £30 Sterling per 1000 samples (tubers)

Region of world	Hectares $(\times 10^3)*$	Mean yield $(t\ ha^{-1})*$	US dollars $(\times 10^3)$
Switzerland	23	41	90
Developed countries			
North America	654	29	2559
West Europe	2370	22	9274
Oceania & others	230	24	900
Developing countries			
Africa	399	7	1561
Latin America	1094	10	4281
Near & Far East	1359	12	5318
Centrally planned economies			
Asia	1648	9	6449
Europe/USSR	10 681	14	41 795
World	18 435	15	72 137

FAO Production Yearbook 1977/79.

The overall costs for the development and application of serology are no longer borne entirely by publicly funded institutes, and research and development work within the public and private sector might be stimulated and improved. The publicly funded research facilities are also relieved of the routine production of serological reagents.

Improvements

The present range of commercially available serological reagents covers only a part of the most requested diagnostic products and there is still room for more diversification. The quality of various reagents can still be improved, especially by using monoclonal antibodies. Experience at Bioreba gained over 8 years comparing monoclonal antibodies with polyclonal antibodies for the detection and diagnosis of potato viruses, has proven their superiority both in production and application. The provision of internationally agreed and quoted reference standards for antigens in diagnostic kits could also considerably improve results. Further refinement of the product specification and of the test protocols would be welcome.

Future applications

The development of reagents for the rapid detection of, as yet uncharacterized, viruses or virus-like, important diseases, or for viruses which are not yet serologically detectable will be an important area for future commercial expansion, both for producers and customers. Farmers, plant nursery owners, dealers in plant material, advisers and phytosanitary officers will expect new, rapid, simple and sensitive field tests, as will laboratories offering diagnostic services. Together they will increase the demand for phytodiagnostic reagents. Extremely sensitive immunoassays will also be needed to measure some epidemiological parameters in order to develop useful forecasting systems for plant diseases. As well as improved serology, there will also be a need to automate laboratory diagnosis.

Conclusions

The commercial application of serology of plant viruses is part of the modern developments in plant biotechnology. Although there is a considerable gap between present and potential volumes, the increasing demand for improved, prophylactic plant protection will lead to the steady evolution of new technology, and of commercial opportunities. It will also validate the worldwide scientific investment which has been made in plant virology.

References

Clark M.F. & Adams A.N. (1977) Characteristics of the microplate method of enzyme-linked immunosorbent assay for the detection of plant viruses. *Journal of General Virology* **34**, 475–83.

Dietzgen R.G. & Sander E. (1982) Monoclonal antibodies against a plant virus. *Archives of Virology* **74**, 197–204.

FAO *The 1977 FAO Production Yearbook*, Vol. 31. FAO, Rome.

Gugerli P. (1982) Improvements of routine tests for potato virus identification. *Proceedings of the Decennial Anniversary Congress of The International Potato Center*, Lima, Peru, pp. 91–2. The International Potato Center, Peru.

Gugerli P. & Fries P. (1983) Characterization of monoclonal antibodies to potato virus Y and their use for virus detection. *Journal of General Virology* **64**, 2471–7.

Voller A., Bartlett A., Bidwell D.E., Clark M.F. & Adams A.N. (1976) The detection of viruses by enzyme-linked immunosorbent assay (ELISA). *Journal of General Virology* **33**, 165–7.

Appendix: Names and addresses of companies supplying biochemicals and equipment

Agdia Inc., 30380, County Road 6, Elkhart, IN 46514, USA

Agri-Diagnostics Associates, 2611 Branch Pike, Cinnaminson, NJ 08077, USA

Aldrich Chemical Company Ltd, The Old Brickyard, New Road, Gillingham, Dorset SP8 4JL, UK

Amersham International plc, Lincoln Place, Green End, Aylesbury, Bucks HP20 2TP, UK

API-bio Mérieux (UK) Ltd, Grafton Way, Basingstoke, Hants RG22 6HY, UK

Applied Biosystems Ltd, Kelvin Close, Birchwood Science Park, Warrington, Cheshire WA3 7PB, UK

ATCC (American Type Culture Collection), 12301 Parklawn Drive, Rockville, Maryland 208, USA

BBL Microbiology Systems, PO Box 243, Cockeysville, MD 21030, USA

BCL (Boehringer Corp. (London) Ltd), Boehringer Mannheim House, Bell Lane, Lewes, East Sussex BN7 1LG, UK

Beckman Instruments (UK) Ltd, Progress Road, Sands Industrial Estate, High Wycombe, Bucks HP12 4JL, UK

Biolog Inc., 3447 Investment Boulevard, Suite 2, Hayward, California 94545, USA

Biometra, Biomedizinische Analytik GmbH, Wagenstieg 5, D-3400 Göttingen, Germany

Bioproducts, FMC Bioproducts Rockland, Maine, USA

Bio-Rad Laboratories Ltd, Bio-Rad House, Marylands Avenue, Hemel Hempstead, Herts HP2 7TD, UK

Bioreba AG, Gempenstrasse 27, Basel, Switzerland 4008

Cambridge BioScience, Newton House, 42 Devonshire Road, Cambridge, Cambs CB1 2BL, UK

Citi-flour Ltd, Connaught Buildings, City University, Northampton Square, London EC1V 0HB, UK

Crescent Dental Manufacturing Co., Lyons, Illinois, USA

Dental Darby Inc., 100 Bank Avenue, Rockville Center, New York, NY 11570, USA

Difco Laboratories Ltd, PO Box 14B, Central Avenue, East Molesey, Surrey KT8 0SE, UK

Dow Chemical Co. Ltd, Lakeside House, Stockley Park, Uxbridge, Middx UB11 1BE, UK

Dupont (UK) Ltd, Wedgewood Way, Stevenage, Herts, UK

General Diagnostics Ltd, Pottery Road, Dun Laoghaire, County Dublin, Ireland

Hewlett Packard Ltd, Heathside Park Road, Cheadle Hume, Stockport, Cheshire SK3 0RB, UK

Hoefer Scientific Instruments, 654 Minnesota Street, PO Box 77387, San Francisco, California 94107, USA

ICN Biomedicals Ltd, Eagle House, Peregrine Business Park, Gomm Road, High Wycombe, Bucks HP13 7DL, UK

Ingenasa, Hermanos Carcia Noblejas 41, Madrid, Spain

Janssen Chimica, Hyde Park House, Cartwright Street, Newton, Hyde, Cheshire SK14 4EH, UK

Jencons (Scientific) Ltd, Cherry Court Way Industrial Estate, Stanbridge Road, Leighton Buzzard, Beds LU7 8UA, UK

Joyce Loebl Ltd, Dukesway, Team Valley, Gateshead, Tyne & Wear NE11 0PZ, UK

Kimble–Kontes, PO Box 729, 1022 Spruce Street, Vineland, New Jersey 08360, USA

Lab M, Topley House, PO Box 19, Bury, Lancs BL9 6AU

Life Technologies Ltd (Gibco-BRL), PO Box 35, Trident House, Paisley PA3 4EF, UK

Loewe Biochemica GmbH, Nordring 38, D-8156 Otterfing, Germany.

Microbial ID Inc., 115 Barksdale Professional Center, Newark, Delaware 19711, USA

Millipore (UK) Ltd, The Boulevard, Blackmoor Lane, Watford, Herts WD1 8YW, UK

New Brunswick Scientific Co., Biosearch European Headquarters, 6 Colonial Way, Watford, Herts WD2 4PT, UK

Oxoid/Unipath Ltd, Wade Road, Basingstoke, Hants RG24 0PW, UK

Pharmacia Ltd, Pharmacia House, 351 Midsummer Boulevard, Central Milton Keynes, Bucks MK9 3YY, UK

Pierce and Warriner (UK) Ltd, 44 Upper Northgate Street, Chester, Cheshire CH1 4EF, UK

PROMEGA, Bumpers Way Industrial Estate, Bristol Road, Chippenham, Wilts SN14 6LH, UK

Sanofi, 40 Avenue Georges V, F 75008, Paris, France

Schleicher and Schuell GmbH, Grimsehlstrasse 23, PO Box 246, D-3352, Einbeck, Germany

Sebia, 23 Rue Maximilien Robespierre, F 92130, Issy les Moulineaux, France

Serotec, 22 Bankside, Station Approach, Kidlington, Oxford, Oxon 0X5 1JE, UK

Sigma Chemical Company Ltd, Fancy Road, Poole, Dorset BH17 7BR, UK

Taylor-Warton Cryogenic Equipment, PO Box 24426, 1505 N. Main Street, Indianapolis, Indiana 46224, USA

Waters Chromatography Division, Millipore (UK) Ltd, The Boulevard, Blackmoor Lane, Watford, Herts WD1 8YW, UK

Whatman Labsales Ltd, St Leonard's Road, 20/20 Maidstone, Kent ME16 0LS, UK

Zymed Laboratories Inc., 25 South Linden Avenue, South San Francisco, California 94080, USA

Index